ENVIRONMENTAL REMEDIATION TECHNOLOGIES, REGULATIONS AND SAFETY

METHANE

EMISSION SOURCES AND REDUCTION STRATEGIES

ENVIRONMENTAL REMEDIATION TECHNOLOGIES, REGULATIONS AND SAFETY

Additional books in this series can be found on Nova's website under the Series tab.

Additional e-books in this series can be found on Nova's website under the e-book tab.

ENVIRONMENTAL REMEDIATION TECHNOLOGIES,
REGULATIONS AND SAFETY

METHANE

EMISSION SOURCES AND REDUCTION STRATEGIES

ANDREA MALONE
EDITOR

nova publishers
New York

NOTICE TO THE READER

The Publisher has taken reasonable care in the preparation of this book, but makes no expressed or implied warranty of any kind and assumes no responsibility for any errors or omissions. No liability is assumed for incidental or consequential damages in connection with or arising out of information contained in this book. The Publisher shall not be liable for any special, consequential, or exemplary damages resulting, in whole or in part, from the readers' use of, or reliance upon, this material. Any parts of this book based on government reports are so indicated and copyright is claimed for those parts to the extent applicable to compilations of such works.

Independent verification should be sought for any data, advice or recommendations contained in this book. In addition, no responsibility is assumed by the publisher for any injury and/or damage to persons or property arising from any methods, products, instructions, ideas or otherwise contained in this publication.

This publication is designed to provide accurate and authoritative information with regard to the subject matter covered herein. It is sold with the clear understanding that the Publisher is not engaged in rendering legal or any other professional services. If legal or any other expert assistance is required, the services of a competent person should be sought. FROM A DECLARATION OF PARTICIPANTS JOINTLY ADOPTED BY A COMMITTEE OF THE AMERICAN BAR ASSOCIATION AND A COMMITTEE OF PUBLISHERS.

Additional color graphics may be available in the e-book version of this book.

Library of Congress Cataloging-in-Publication Data

ISBN: 978-1-63482-526-9

Published by Nova Science Publishers, Inc. † New York

CONTENTS

PREFACE

On June 25, 2013, President Obama announced a national "Climate Action Plan" (CAP) to reduce emissions of carbon dioxide and other greenhouse gases, as well as to encourage adaptation to expected climate change. One of the more significant initiatives within the CAP focused on the control of methane emissions, a potent short-lived climate pollutant. It called for the U.S. Environmental Protection Agency (EPA) and the Departments of Agriculture, Energy, Interior, Labor, and Transportation to develop a comprehensive interagency "Strategy to Reduce Methane Emissions" (Strategy). The Strategy commits to new steps to cut emissions through both voluntary actions and proposed rulemaking, and outlines the Administration's efforts to improve the measurement and assessment of these emissions. This book's objective is to discuss emission sources of methane, and to examine strategies of reducing greenhouse gases.

Chapter 1 - Methane is the world's simplest hydrocarbon, with a chemical formula CH4 (one atom of carbon and four atoms of hydrogen). It is gaseous under normal atmospheric conditions, and is commonly produced through the decomposition of organic materials in the absence of oxygen. It is released into the atmosphere by natural sources such as wetlands, oceans, sediments, termites, volcanoes, and wildfires,[1] as well as human activities such as oil and natural gas systems, coal mines, landfills, wastewater treatment facilities, and the raising of livestock.

This report examines the many facets of methane: from commodity to coproduct to by-product to waste. It begins with a survey of past and present attempts by Congress and the executive branch to address methane emissions for the purposes of energy policy and pollution control. It then provides a general overview of methane before focusing on specific sectors of the economy in order to (1) characterize different sources of methane and the data available on their emissions; (2) discuss current practices, opportunities, and challenges for emission controls; and (3) outline recent initiatives proposed by Congress and the Administration.

Chapter 2 - Methane is both a potent greenhouse gas and a valuable source of energy. In the Climate Action Plan, President Obama directed the Administration to develop a comprehensive, interagency strategy to reduce methane emissions. In March 2014, the White House released the *Climate Action Plan - Strategy to Reduce Methane Emissions*. As part of the Strategy, the U.S. Department of Agriculture (USDA), the U.S. Environmental Protection Agency (EPA) and the U.S. Department of Energy (DOE) committed to work with industry leaders to formulate a biogas roadmap.

This *Biogas Opportunities Roadmap* builds on progress made to date to identify voluntary actions that can be taken to reduce methane emissions through the use of biogas systems and outlines strategies to overcome barriers to a robust biogas industry in the United States. It supports the U.S. dairy industry's voluntary 2008 goal to reduce its greenhouse gas emissions by 25% by 2020. This goal was a driver behind a partnership forged between the Dairy Industry and USDA in December 2009 and renewed in May 2013. As part of the 2013 renewal, the dairy industry also requested that USDA create a voluntary biogas roadmap to support this goal.

Biogas is a proven source of energy used in the United States and around the world for decades. The United States currently has more than 2,000 sites producing biogas. The Roadmap found that with the proper support, more than 11,000 additional biogas systems could be deployed in the United States. If fully realized, these biogas systems could produce enough energy to power more than 3 million American homes and reduce methane emissions equivalent to 4 to 54 million metric tons of greenhouse gas emissions in 2030, the annual emissions of between 800,000 and 11 million passenger vehicles.

In order to realize these opportunities, the Roadmap identifies actions the Federal government will take to increase the use of biogas to meet our renewable energy goals, strengthen the economy, and reduce methane emissions exclusively through voluntary actions.

These include:

- *Promoting Biogas Utilization through Existing Agency Programs*: USDA, DOE, and EPA will use their existing programs as a vehicle to enhance the utilization of biogas systems in the U.S by ensuring that existing criteria for technical and financial assistance considers the benefits of biogas systems, leveraging over $10 million in research funding to enhance the economic viability and benefits of biogas systems and co-products, and strengthening programs that support the use of biogas for clean energy, transportation fuel, renewable chemicals and biobased products.

- *Fostering Investment in Biogas Systems*: To help overcome financial barriers to the widespread investment in biogas systems, USDA will lead efforts to improve the collection and analysis of industry financial and technical data needed to track the performance of anaerobic digesters, evaluate current loan and grant programs for opportunities to broaden the financing options available for biogas systems, and review Federal procurement guidelines to ensure that products of biogas systems are eligible for and promoted by applicable government procurement programs.
- *Strengthening Markets for Biogas Systems and System Products*: To strengthen U.S. markets for renewable energy and value-added non-energy products from biogas systems, USDA, DOE, and EPA will review opportunities to overcome barriers to integrating biogas into electricity and renewable natural gas markets, for example, though modernizing existing Federal incentives provided for renewable energy generation. USDA, EPA, and DOE will also drive the creation of tools to help industry broaden the market development for energy and non-energy biogas systems products.
- *Improving Communication and Coordination*: In order to implement the strategies laid out in this document and promote strong coordination and messaging across Federal agencies, USDA will establish a Biogas Opportunities Roadmap Working Group that will include participation from DOE and EPA, as well as the dairy and biogas industries. The Working Group will collaborate with industry to publish a progress report in August 2015, which identifies and prioritizes policies and technology opportunities to expand the biogas industry and reduce greenhouse gas emissions.

The emissions intensity of the production of meat and milk in the United States is already much lower today than it was even a few decades ago. Due to improvements in production efficiency, it's amongst the lowest in the world. Enhancing the deployment of cost-effective technology to utilize biogas can increase revenues and reduce emissions, providing another "win-win" for farmers, communities, and the nation.

Chapter 3 - Reducing methane emissions is a powerful way to take action on climate change; and putting methane to use can support local economies with a source of clean energy that generates revenue, spurs investment, improves safety, and leads to cleaner air. That is why in his Climate Action Plan, President Obama directed the administration to develop a comprehensive, interagency strategy to cut methane emissions. This document lays out that strategy – summarizing the sources of methane and trends in emissions; setting forth a plan to reduce both domestic and international methane emissions through incentive-based programs and the Administration's existing authorities; and, outlining the Administration's efforts to improve measurement of these emissions. This strategy also highlights examples of technologies and industry-led best practices that are already helping to cut methane emissions.

Chapter 4 - Methane gas is a valuable energy resource and the leading anthropogenic contributor to global warming after carbon dioxide. Atmospheric methane concentrations have doubled over the last 200 years and continue to rise, although the rate of increase is slowing. By mass, methane has 21 times the global warming potential of carbon dioxide over a 100-year time frame. Methane accounts for 10 percent of U.S. greenhouse gas emissions (excluding sinks) and reducing these emissions is a key goal of the U.S. Climate Change Action Plan.

The major sources of anthropogenic methane emissions in the U.S. are landfills, agriculture (livestock enteric fermentation and manure management), natural gas and oil systems, and coal mines. Smaller sources in the U.S. include rice cultivation, wastewater treatment, and others. Unlike other greenhouse gases, methane can be used to produce energy since it is the major component (95 percent) of natural gas. Consequently, for many methane sources, opportunities exist to reduce emissions cost-effectively or at low cost by capturing the methane and using it as fuel.

This report has two objectives. First, it presents the U.S. Environmental Protection Agency's (EPA's) baseline forecast of methane emissions from the major anthropogenic sources in the U.S., and EPA's cost estimates of reducing these emissions. Emission estimates are given for 1990 through 1997 with projections for 2000 to 2020. The cost analysis is for 2000, 2010, and 2020. Second, this report provides a transparent methodology for the calculation of emission estimates and reduction costs, thereby enabling analysts to replicate these results or use the approaches described herein to conduct similar analyses for other countries.

In: Methane
Editor: Andrea Malone

ISBN: 978-1-63482-526-9

Chapter 1

METHANE: AN INTRODUCTION TO EMISSION SOURCES AND REDUCTION STRATEGIES*

Richard K. Lattanzio, Kelsi Bracmort, Anthony Andrews, James E. McCarthy and Lynn J. Cunningham

SUMMARY

The Obama Administration's Strategy to Reduce Methane Emissions

On June 25, 2013, President Obama announced a national "Climate Action Plan" (CAP) to reduce emissions of carbon dioxide (CO_2) and other greenhouse gases (GHGs), as well as to encourage adaptation to expected climate change. One of the more significant initiatives within the CAP focused on the control of methane emissions, a potent short-lived climate pollutant. It called for the U.S. Environmental Protection Agency (EPA) and the Departments of Agriculture, Energy, Interior, Labor, and Transportation to develop a comprehensive interagency "Strategy to Reduce Methane Emissions" (Strategy). The Strategy, released on March 28, 2014, commits to

---

* This is an edited, reformatted and augmented version of a Congressional Research Service publication, No. R43860, dated January 14, 2015.

new steps to cut emissions through both voluntary actions and proposed rulemaking, and outlines the Administration's efforts to improve the measurement and assessment of these emissions.

Perspectives on the Strategy

Many of the affected industries (including some in the agriculture, fossil energy, and waste management sectors) have raised concerns over increased controls. They argue that further regulation of emissions would be either untenable from an economic standpoint or ineffective at providing significant health and environmental benefits. They contend that industries are already doing everything feasible to capture and reuse methane emissions (for requisite safety and economic reasons), and that state and local authorities—who share a closer understanding of the industries' specific circumstances—are best equipped to oversee and enforce emission reduction efforts within their jurisdictions. Some U.S. lawmakers support these viewpoints, and have proposed legislation to roll back the federal initiatives.

Health and environmental advocates, however, contend that the Strategy falls short. They argue that methane emissions can jeopardize worker safety, lead to ground-level ozone formation (commonly referred to as "smog"), and act as a potent GHG. Recent events in the United States (e.g., the rise in domestic oil and natural gas production, the encroachment of domestic oil and natural gas production on new or more populated areas, and the revitalization of the petrochemical manufacturing sector) have led these stakeholders to suggest the need for more enforceable standards. Some U.S. lawmakers agree, and they have proposed new controls or pushed for federal agencies to more fully regulate methane emissions.

The Role of Methane

Behind it all is methane—the world's simplest hydrocarbon and the primary component of natural gas. It is released into the atmosphere by both natural sources (such as wetlands and wildfires) and human activities (such as oil and natural gas systems, coal mines, landfills, and the raising of livestock). When captured, methane can be used as either a fuel or a chemical feedstock, with many advantages over other fossil fuels (e.g., it is more versatile and less polluting, and provides energy security benefits). Its dual nature as both a pollutant and a commodity makes efforts to control emissions potentially beneficial to both the economy and the environment.

For these reasons, the federal government has sought policies to help reduce, capture, and reuse methane emissions as far back as the 1970s. Whether strategies to control emissions are effective and cost-efficient for a given industry may depend upon a number of factors including the nature and extent of the emissions, the technology available for capture, and the market price for the recovered products. In this way, the cost-benefit considerations are similar to those of energy efficiency efforts, wherein high up-front investments and other market barriers, if confronted by producers, may have the potential to be offset over time. Recent federal policies have included a variety of funding programs for research and technology development, as well as voluntary programs and tax incentives for industry. Currently, methane emissions are addressed directly by two federal rules: one on new municipal landfills and another on federal oil and gas leases. The Obama Administration's recent Strategy—as well as a variety of recent proposals in Congress— attests to the continued interest in an appropriate policy response to the issue of methane emissions.

INTRODUCTION

Methane is the world's simplest hydrocarbon, with a chemical formula CH_4 (one atom of carbon and four atoms of hydrogen). It is gaseous under normal atmospheric conditions, and is commonly produced through the decomposition of organic materials in the absence of oxygen. It is released into the atmosphere by natural sources such as wetlands, oceans, sediments, termites, volcanoes, and wildfires,[1] as well as human activities such as oil and natural gas systems, coal mines, landfills, wastewater treatment facilities, and the raising of livestock.

Methane, when captured, can be used as either a fuel or a chemical feedstock. When used as a fuel—for example, methane is the primary component of natural gas[2]—it has many advantages over other hydrocarbons (e.g., oil and coal). Methane is more versatile; it can heat homes, fuel stoves, run vehicles, fire power plants, and, when liquefied, be exported to support the energy needs of U.S. allies and trading partners. Methane is cleaner-burning; it emits, on average, about half as much CO_2 as coal and one-quarter less than oil when consumed in a typical electric utility plant.[3] Further, its combustion emits no mercury (a persistent, bioaccumulative neurotoxin), virtually no particulate matter, and less sulfur dioxide and nitrogen oxide, on average, than either coal or oil. Recent expansion in natural gas production, primarily as a result of improved technologies (e.g., hydraulic fracturing and directional drilling)[4] used on unconventional resources (e.g., shale, tight sands, and coal-bed methane),[5] has made methane an increasingly significant component in the energy supply and security of the United States.

When used as a chemical feedstock, methane is a manufacturing component for a variety of household and industrial products including plastic, fertilizer, antifreeze, and fabrics. Abundant and economical supplies of methane may serve arguably to reinvigorate the U.S. petrochemical sector, bringing manufacturing industries back on shore and aiding in the creation of domestic jobs and economic development. For these reasons, many in both the public and private sector have advocated for the increased production and use of methane (via natural gas extraction or other capture technologies), and have hailed it as a potential "cost-effective bridge" to a less polluting and lower GHG-intensive economy.[6] This position has been supported by many members of Congress as well as the Obama Administration.[7]

Methane, however, when released or allowed to escape into the atmosphere (commonly referred to as "vented" and "fugitive" emissions, respectively), has adverse impacts on human health, safety, and the environment. The U.S. Occupational Safety and Health Administration (OSHA) lists methane as both an asphyxiant and an explosive, as increased concentrations in local settings can jeopardize worker safety.[8] Further, the U.S. Environmental Protection Agency (EPA) classifies methane as both a precursor to ground-level ozone formation[9] (commonly referred to as "smog") and a potent greenhouse gas (GHG), albeit with a shorter atmospheric life than CO_2.[10] Methane's effect on climate change is up to 34 times greater than that of CO_2 when considered over a 100-year time period, and even greater when considered over the first 20 years after it is emitted.[11] An increase in emissions may counteract some of the environmental benefits that the U.S. economy has to gain by switching from coal or oil to natural gas and other sources of methane. For these reasons, some stakeholders, including some Members of Congress, have called for increased controls on methane emissions in several sectors of the economy including oil and natural gas production, coal mining, industrial processes, and agriculture.

In many cases, efforts to control air pollution can compete against the economic considerations of the affected industries. However, in methane's case, its dual nature as both a commodity and a pollutant provides a unique set of incentives. Under certain conditions, the value of fugitive methane and other by-products that can be recovered and sold at market may be able to offset the cost of their capture. Further, the value of these recovered products during oil and gas extraction would contribute to increased royalty payments to state and federal governments.

The difficulty, however, is that methane emissions are not always easy to capture. Methane, unlike some other pollutants (e.g., sulfur dioxide or CO_2), is not commonly emitted in a concentrated stream from industrial processes. Rather, it is released into the atmosphere through dispersion, leaks, vents, accidents, and ruptures. In this way, methane emissions are most similar to those of volatile organic compounds (VOCs), both in manner and control.[12] Efforts to capture or abate these emissions are generally more difficult and costly than for other pollutants. Whether or not recovery of methane is profitable for producers may depend upon a number of factors including the nature and extent of the release, the technology available for capture, and the market price for the recovered products. In this way, the cost-benefit consideration of methane capture becomes very similar to that of energy efficiency efforts, wherein high up-front investments and other market barriers, if confronted by producers, may have the potential to be offset over time.

This report examines the many facets of methane: from commodity to coproduct to by-product to waste. It begins with a survey of past and present attempts by Congress and the executive branch to address methane emissions for the purposes of energy policy and pollution control. It then provides a general overview of methane before focusing on specific sectors of the economy in order to (1) characterize different sources of methane and the data available on their emissions; (2) discuss current practices, opportunities, and challenges for emission controls; and (3) outline recent initiatives proposed by Congress and the Administration.

ISSUES FOR CONGRESS

Through the years, the federal government has sought policies to control methane emissions for a variety of economic, environmental, and public health and safety reasons. Some justifications for federal involvement have included the following:

1) promoting domestic energy production and energy independence,
2) protecting the property rights of mineral owners (including federal resources and associated royalties to the American taxpayer),
3) assuring the operational safety of employees who work with or near significant emission sources, and
4) safeguarding the general population from air pollution that may reasonably be anticipated to endanger public health or welfare.

Initially, policies to capture methane emissions were motivated in part by the Organization of Arab Petroleum Exporting Countries oil embargo of 1973 and the subsequent calls for U.S. energy independence. During this time, the United States saw natural gas and other sources of methane as a potential alternative to imported crude oil. Efforts to incentivize the capture of methane and utilize it as an alternative fuel were proposed by both Congress and the Administration across the full range of commercial sectors. They included a variety of funding programs for research and technology development, voluntary guidelines and tax incentives for industry, and/or rules for mineral rights lessees on federal lands.

As an understanding of methane's role in ozone formation and climate change grew during the 1990s, some state and federal authorities turned their attention to reducing methane emissions as a form of pollution control. Once again, the key policy tools used for pollution abatement took the form of voluntary guidelines and tax incentives. However, in a few instances, where reductions in methane emissions could serve the co-benefit of aiding in the reduction of other pollutants, regulatory emission standards were proposed and/or promulgated. In the 2000s, as Congress considered comprehensive market-based strategies to reduce GHG emissions across the entire U.S. economy, more innovative proposals for methane reduction became prevalent. Methane capture was commonly suggested as an "offset" credit for higher GHG-emitting industries, as the net costs of reducing methane emissions, in some instances, could be more favorable than directly controlling for CO_2 emissions.

Recent events in the United States (e.g., the rise in domestic oil and natural gas production, its encroachment on new or more populated areas, and the revitalization of the petrochemical manufacturing sector) have led some stakeholders to suggest the need for more enforceable standards. At the state level, Colorado, Wyoming, Ohio, and California have recently promulgated or proposed rules to control for methane emissions from their oil and gas sectors.[13] At the federal level, two methane-emitting source categories are addressed directly by regulations. They include (1) EPA's 1996 standards on municipal landfills,[14] and (2) the Bureau of Land Management's (BLM's) 1980 notice on venting and flaring for oil and gas leases on federal lands.[15] However, many emission sources in the oil and gas industry, as well as many activities in the agricultural and waste management sectors, remain uncovered by any regulatory standard. For this reason, some U.S. lawmakers have proposed controls and/or have pushed for federal agencies to more fully regulate methane emissions.

EPA has the authority to regulate methane emissions as both an ozone precursor and a GHG under the Clean Air Act (CAA).[16] Currently, EPA has no standards in place to regulate methane as an ozone precursor, and it has shown a disinclination for doing so in the past.[17] The agency's authority to regulate methane as a GHG has been upheld by the Supreme Court's 2007 decision in *Massachusetts v. EPA*[18] which determined that GHGs fall under the definition of "air pollutant" as used in the CAA. Following this decision, EPA determined that six GHGs, including methane, endangered public health and welfare,[19] and issued several rules to control GHGs focused primarily on CO_2.[20] The Administration has been pressured by many health and environmental organizations to promulgate performance standards specific to methane emissions. However, to this point, EPA has not advanced such standards. The agency maintains that existing rules set on other air pollutants commonly co-emitted with methane (e.g., VOCs) as well as its voluntary programs with industry have adequately returned cost-effective reductions in emissions.[21] In addition to EPA, the Departments of Agriculture, Energy, Interior, Labor, and Transportation have some authorities to monitor, give guidance for, and make rules to control for methane emissions. Current and proposed initiatives—including the Obama Administration's 2014 "Strategy to Reduce Methane Emissions"—are discussed in further detail in the remainder of this report.

Many affected industries—specifically those in the energy and the agricultural sectors—have raised concerns over increased controls. They argue that further regulation would be either untenable from an economic standpoint or ineffective at providing significant health and environmental benefits. They contend that industries are already doing everything feasible to capture and reuse methane emissions (for requisite safety and economic reasons), and that state and local authorities—who share a closer understanding of an industry's specific circumstances— are best equipped to oversee and enforce any emission reduction efforts within their jurisdictions. Some U.S. lawmakers have supported these viewpoints.

Efforts by the federal government to incentivize the reduction, capture, and reuse of methane emissions are summarized in the following two sections: "Legislative Initiatives" and "Administrative Initiatives." Further, Table A-1 of Appendix A provides a detailed list of recent congressional proposals both in support of and in opposition to increased methane emission controls. Finally, Appendix B provides a selected chronology of recent executive branch initiatives.

Legislative Initiatives

The U.S. Congress has pursued policies in support of methane reduction since the 1970s. Legislation aimed at capturing methane emissions from agricultural activities and promoting the use of the recovered gas dates back, at least, to the 94th Congress.[22] Similar bills targeting emissions from coal mines and municipal landfills were introduced in the 96th and 97th Congresses, respectively.[23] These efforts often promoted methane as an alternative fuel source, specifically as a replacement for imported crude oil. Legislation addressing methane's role as an air pollutant (e.g., as a GHG) reaches back to the 101st Congress, wherein several bills were introduced with specific methane control provisions. These included one in 1989 by then Senator Al Gore to analyze "the contribution of methane to global climate change, the sources and sinks of methane, and the methods of controlling emissions of methane."[24] A similar set of studies was codified by the Clean Air Act Amendments of 1990, which required EPA to report on the "activities, substances, processes, or combinations thereof that could reduce methane emissions and that are economically and technologically justified."[25] Methane reduction was also included as a qualifying activity in market-based GHG control proposals as far back as the 101st Congress.[26]

Recent congressional interest continues to focus on methane's role as a GHG, with legislative efforts aimed at both supporting EPA's authority to regulate methane emissions as well as revoking it. Recent bills and amendments have proposed several different policy tools as strategies for reduction. They include (1) providing economic incentives (e.g., through tax benefits) for activities that capture and use fugitive gas (e.g., H.R. 860, the Biogas Investment Tax Credit Act of 2013), (2) authorizing the Administration or a specific agency to investigate or directly regulate methane emissions (e.g., H.Amdt. 507 to H.R. 2728 sought to allow the Secretary of the Interior to issue regulations to reduce methane emissions from oil and gas operations on federal and Indian lands), and (3) providing a market-based mechanism (e.g., fee) to incentivize methane reduction (e.g., S. 332, the Climate Protection Act of 2013).

Conversely, many bills in recent Congresses have also aimed to remove the executive branch's authority to regulate methane emissions based predominantly on arguments for economic growth and employment. Some examples of the most recent efforts include (1) amending the CAA to remove "methane" and other GHGs from the definition of "air pollutant" (e.g., H.R. 3895, the Energy Exploration and Production to Achieve National Demand Act [of 2014]), and (2) prohibiting appropriated funds from being used by agencies to regulate methane (e.g., H.R. 621, the Ensuring Affordable Energy Act [of 2013]).

For a selected list of recent bills and amendments that address methane, see Table A-1 of Appendix A.

Administrative Initiatives

Historically, many of the methane control initiatives managed by the federal government have taken the form of either research and development programs or voluntary public-private partnerships with industry. Federal research and development programs have provided funding for new technologies to enable more cost-effective emission reductions across various sectors of the economy. Offices that have provided financial and technical assistance include the Department of Agriculture (USDA) Conservation Innovation Grants, Environmental Quality Incentive Program, Rural Energy for America Program, Bioenergy Program for Advanced Biofuels, and Biorefinery Assistance Program; the Department of Energy (DOE) Office of Fossil Fuels, Office of Energy Policy and Systems Analysis, and Section 1703 Loan Guarantee Program; the Department of Labor (DOL) Mine Safety and Health Administration; and the Department of Transportation (DOT) Pipeline and Hazardous Materials Safety Administration, as well as the EPA Office of Air and Radiation and the Department of the Interior (DOI) BLM.

Similarly, voluntary partnerships managed by federal agencies have aimed to leverage the resources of the federal government to assist the private sector in overcoming the economic barriers to methane capture. They include the EPA's Natural Gas STAR Program and the Coalbed Methane Outreach Program for the energy sector, EPA/USDA's AgSTAR Program for the agricultural sector, EPA's Landfill Methane Outreach Program for the waste sector, and EPA's Global Methane Initiative for international activities.[27] The goals of these programs are to (1) raise awareness of emission levels and the value of lost fuel, (2) provide information and training on new technologies and practices, and (3) discuss the barriers embedded in traditional operations, limited infrastructure, and uncertain investment climates. As with many voluntary initiatives, these programs have returned mixed results.[28]

The Obama Administration's Strategy to Reduce Methane Emissions

On June 25, 2013, President Obama refocused his Administration's efforts to address GHG emissions with the release of the "Climate Action Plan" (CAP).[29] Federal activities in support of methane emission reductions became one of the cornerstones of the CAP. During its presentation, the President stated that "curbing emissions of methane is critical to our overall effort to address global climate change." Many stakeholders have suggested that the Administration's recent GHG reduction targets, offered under the U.S. commitments to the United Nations Framework Convention on Climate Change, would be unattainable without significant methane controls. The CAP set guidelines for EPA and the Departments of Agriculture, Energy, Interior, Labor, and Transportation to develop a comprehensive interagency methane strategy,[30] which was released on March 28, 2014, under the title "Strategy to Reduce Methane Emissions" (Strategy).[31]

Key initiatives of the Strategy include the following:

1) **Agriculture.** A joint USDA, EPA, and DOE "Biogas Roadmap" outlining voluntary strategies to accelerate adoption of methane digesters and other cost-effective technologies to reduce U.S. dairy sector GHG emissions by 25% by 2020 (released on August 1, 2014).[32]
2) **Petroleum and Natural Gas.**
   - An EPA assessment of several potentially significant sources of methane and other emissions from the oil and gas sector through a series of technical white papers[33] and a determination on how best to pursue further methane reductions from these sources. The White House announced on January 14, 2015, that these steps would include (1) a proposal to build on a set of 2012 New Source Performance Standards (NSPS) for VOC emissions to address new and modified activities and equipment in the sector uncovered by the previous rule (scheduled for release in the summer of 2015), (2) extending VOC reduction requirements to

existing oil and gas sources in ozone nonattainment areas and states in the Ozone Transport Region, and (3) expanding voluntary efforts under the Natural Gas STAR program.[34]

- A BLM proposal to update standards to reduce venting and flaring from oil and gas production on public lands (scheduled for release in April 2015).
- A Department of Transportation Pipeline and Hazardous Materials Safety Administration (PHMSA) proposal for natural gas pipeline safety standards (scheduled for release in 2015).
- DOE-convened roundtables, as part of the Quadrennial Energy Review, to identify "downstream" methane reduction opportunities (the summary of which was released on July 29, 2014).[35]

3) **Coal Mines.** A BLM Advance Notice of Proposed Rulemaking (ANPRM) to gather public input on the development of a program for the capture and sale, or disposal, of waste mine methane on lands leased by the federal government (released on April 28, 2014).[36]
4) **Landfills.** An EPA proposal to update standards to reduce methane from new landfills, and to take public comment on whether to update standards for existing landfills (released on July 17, 2014).[37]
5) **Improving Methane Measurement.** Data quality improvement, including developing new measurement technologies, addressing areas of higher uncertainty in bottom-up inventories, and enhancing top-down modeling and monitoring based on direct measurement of atmospheric concentrations.

These initiatives are summarized in greater detail, by sector, in the remainder of this report. For a selected chronology of executive branch initiatives related to the White House's Strategy, see Appendix B.

METHANE: A PRIMER

Methane is both a precursor to ground-level ozone formation and a potent GHG. As a precursor to ground-level ozone formation, methane reacts with nitrogen oxides in the presence of sunlight to form what is commonly referred to as smog. Methane, however, is generally less reactive than other hydrocarbons. For this reason—and at this time—EPA has excluded it from the definition of regulated hydrocarbons called volatile organic compounds (VOCs).[38]

As a GHG, methane emitted into the atmosphere absorbs terrestrial infrared radiation, which contributes to increased global warming and continuing climate change. According to the Intergovernmental Panel on Climate Change (IPCC) *Fifth Assessment Report 2013* (AR5), in 2011, methane concentrations in the atmosphere exceeded preindustrial levels by 150%. Further, they contributed about 16% to global warming due to anthropogenic GHG sources, making methane the second-leading climate forcer after CO2 globally.[39] While the perturbation lifetime for methane is only 12 years (compared to CO2's, which is considerably longer, and does not undergo a simple decline over a single predictable timescale), its immediate impacts are significantly greater (see Text Box). For this reason, it is commonly characterized as a "short-lived climate forcer," along with black carbon and various hydrofluorocarbons (HFCs).

Global Warming Potential

The climate change impacts of methane are commonly compared to those of CO_2 through the use of an index referred to as "global warming potential" (GWP): a measure of the total energy that a gas absorbs over a particular period of time compared to CO_2. Key factors affecting the GWP of any given gas include its average atmospheric lifetime and the ability of that molecule to trap heat. According to the current metrics used by EPA, the same amount of methane emissions by mass is approximately 25 times more potent than CO_2 emissions when averaged over a 100- year time horizon.[40] Further, methane chemically reacts in the atmosphere to produce other climate warming gases— for example, ozone in the troposphere and water in the stratosphere. An estimate of the warming effects of these product gases is included in the GWP of 25. However, these reactions also indirectly affect aerosols in the atmosphere, likely further enhancing the warming effect of methane.[41]

As stated, methane reacts with other chemicals in the atmosphere and dissipates. Thus, while methane is a highly potent GHG for a short period after its initial release, its capacity to trap heat dissipates after approximately 12 years. By comparison, CO_2's is considerably longer, and does not undergo a simple decline over a single predictable timescale.

Instead, the excess atmospheric carbon from CO_2 emissions mixes into the oceans and biosphere (e.g., plants) over a period of a few hundred years, and then it is slowly removed over hundreds of thousands of years as it is gradually incorporated into carbonate rocks.

As recently as November 2013, EPA reported GWP values for methane that were accepted by Parties to the United Nations Framework Convention on Climate Change (UNFCCC) as they were presented in the IPCC *Second Assessment Report 1995* (SAR). The SAR lists methane's GWP as 21 over a 100-year time horizon. EPA's most recent *Inventory of U.S. Greenhouse Gas Emissions and Sinks: 1990-2012*, released in April 2014, uses the SAR GWP of 21 (and, by extension, all of the data and graphics in this report use the SAR GWP of 21). EPA has recently adopted GWP values for methane that were accepted by Parties to the UNFCCC as they were presented in the IPCC *Fourth Assessment Report 2007* (AR4).[42] The AR4 lists methane's GWP as 25 and 72 over a 100-year and a 20-year time horizon, respectively. EPA's 2015 Inventory will employ these GWPs. Accordingly, due to this reevaluation of climate impacts, methane's comparative role as a GHG will increase by approximately 20% under the new reporting. The IPCC *Fifth Assessment Report 2013* (AR5), released in September 2013, lists methane's GWP as 28 and 84 over a 100- year and a 20-year time horizon, respectively, but these values have not yet been accepted officially by Parties to the UNFCCC. Further, the AR5 reports methane's GWP inclusive of methane's indirect effects on aerosols as 34 and 86 over a 100-year and a 20-year time horizon, respectively.

Emissions

According to EPA, methane is the second-most prevalent GHG emitted by the United States (behind CO2), and in 2012 it accounted for 567.3 million metric tons of CO2 equivalent, or about 9% of all domestically produced emissions from human activities.[43] Some academic studies have put these emissions even higher.[44] Of the total, nearly 40% was emitted from sources in the energy production sector, a third from sources in the agricultural sector, and a fifth from sources in the waste management sector (see Figure 1).

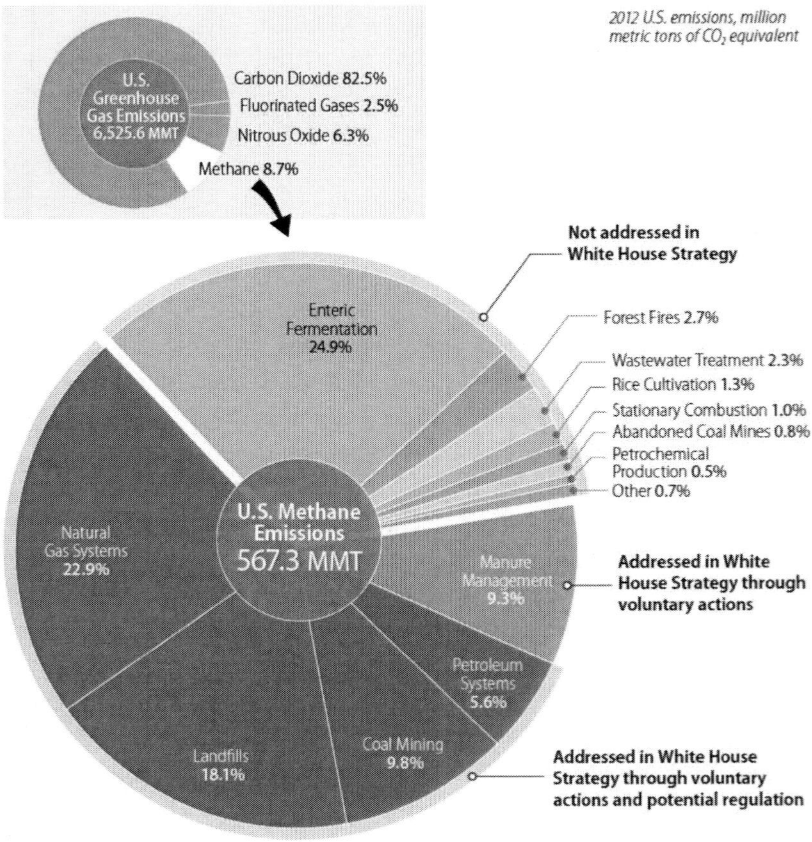

Source: Congressional Research Service, with data from the U.S. Environmental Protection Agency, *Inventory of U.S. Greenhouse Gas Emissions and Sinks: 1990-2012*, Washington, DC, EPA 430-R-14-003, April 15, 2014.

Figure 1. U.S. Methane Emissions: Sources.

Historical Trends

Between 1990 and 2012, methane emissions in the United States decreased by almost 11%. During this time period, emissions have increased from sources associated with agricultural activities, while emissions have decreased from sources associated with waste management and the exploration and production of natural gas and petroleum products (see Figure 2). Comparatively, the source categories for natural gas systems and landfills have seen the most notable reductions over the past 20 years, and manure management the most notable increase, but many other subcategories have seen little to no change (see Figure 3).

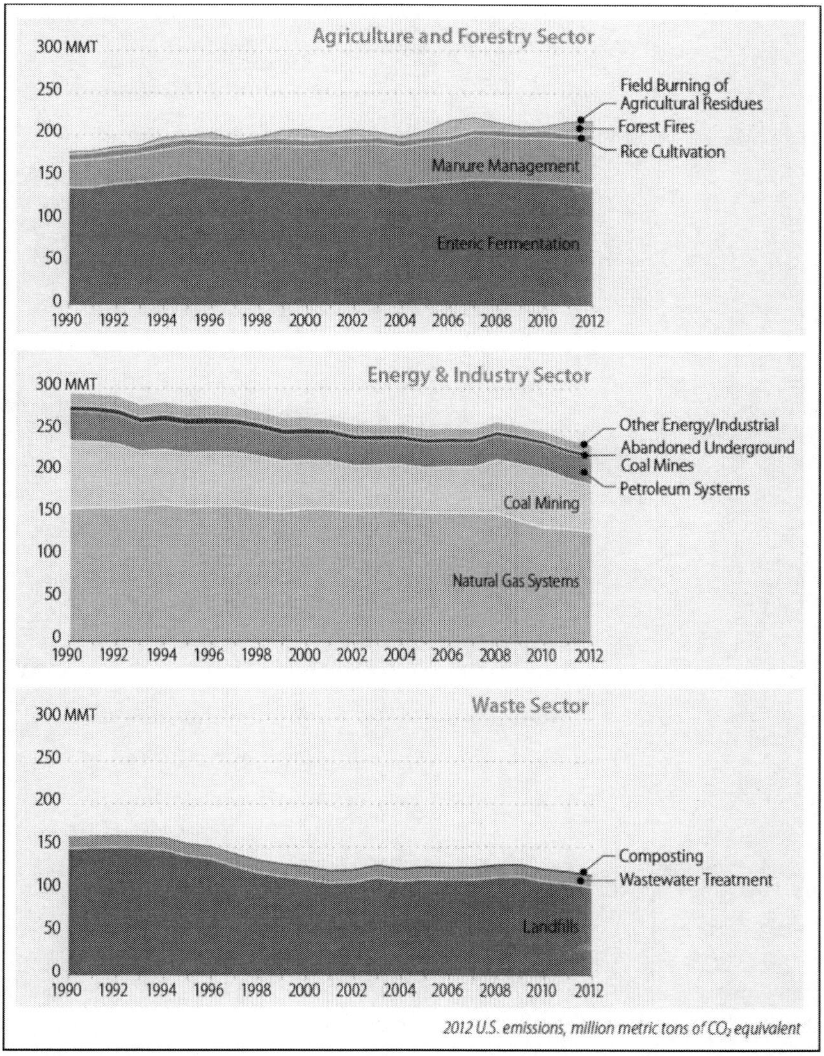

Source: Congressional Research Service, with data from the U.S. Environmental Protection Agency, *Inventory of U.S. Greenhouse Gas Emissions and Sinks: 1990-2012*, Washington, DC, EPA 430-R-14-003, April 15, 2014.

Figure 2. U.S. Methane Emissions: Historical Trends by Source Sector.

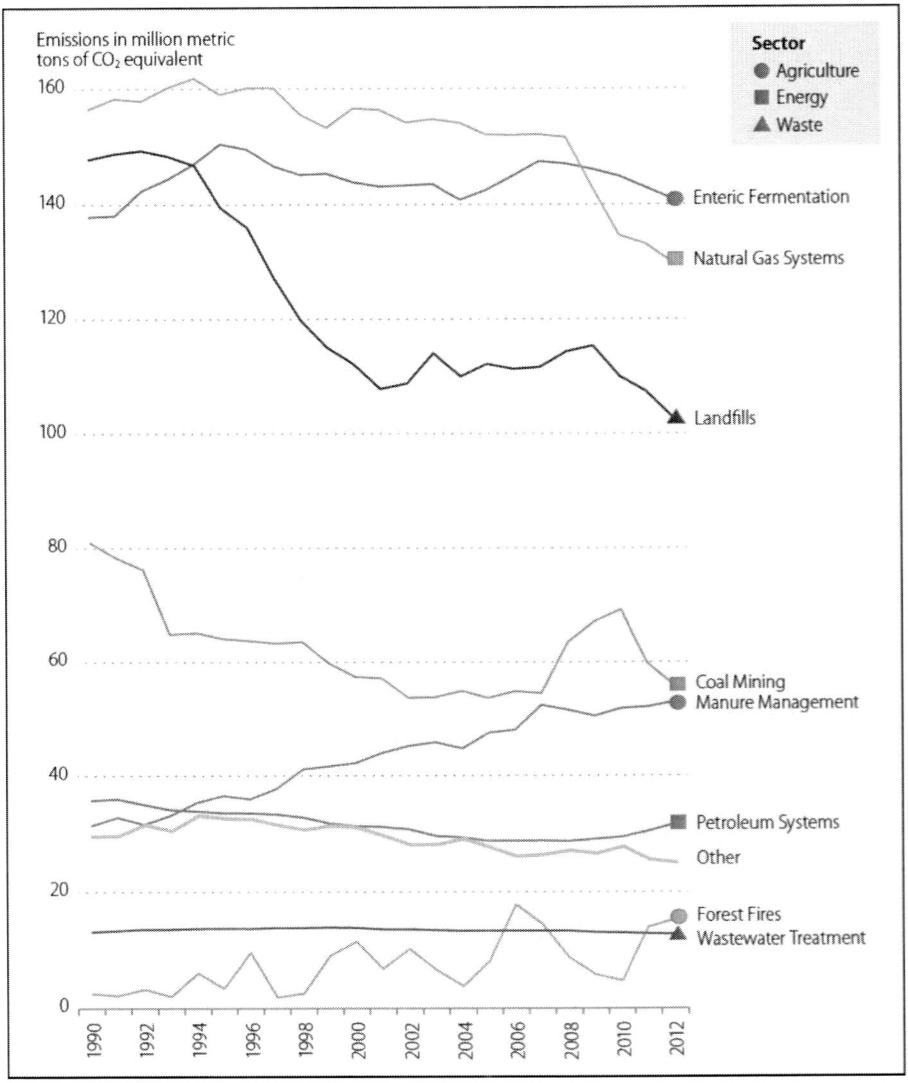

Source: Congressional Research Service, with data from the U.S. Environmental Protection Agency, *Inventory of U.S. Greenhouse Gas Emissions and Sinks: 1990-2012*, Washington, DC, EPA 430-R-14-003, April 15, 2014.

Note: "Other" sources include rice cultivation, stationary combustion, abandoned coal mines, petrochemical production, composting, iron, steel and coke production, and the burning of agricultural residue.

Figure 3. U.S. Methane Emissions: Historical Trends by Source Category.

Source Sectors and Mitigation Activities

Emissions of methane can be categorized into three broad source categories: agriculture, energy and industrial processes, and waste management. The following section reviews each of these categories, and

- characterizes the major sources in the sector and the respective data on emissions,[45]
- discusses current practices, opportunities, and challenges for emission control,
- summarizes current and applicable government programs and program performance data, and
- outlines the proposed initiatives in the White House's recent Strategy.

Agriculture Sector[46]

Agricultural sources of methane emissions include the following:

- **Enteric Fermentation.** Methane is produced as part of normal digestive processes in animals, which is more so an issue with ruminant livestock (e.g., cattle). Microbes that reside in the animal's digestive system ferment food consumed by the animal and produce methane as a by-product, which can be eructated (i.e., belching or flatulence) by the animal.
- **Manure Management.** Methane is produced from manure management systems, primarily liquid and slurry systems. The treatment and storage of livestock manure can produce methane through its anaerobic decomposition.
- **Rice Cultivation.** Methane is produced from the anaerobic environment resulting from flooded fields used for rice cultivation. Decomposition of organic material gradually depletes most of the oxygen present in the soil, causing anaerobic soil conditions.
- **Field Burning of Agricultural Residues.** Methane is emitted from the field burning of agricultural residues, which is done usually for disposal purposes. Field burning of agricultural residues occurs more frequently in some parts of the United States, and is regulated or monitored depending on state and local law. Internationally, slash-and-burn agriculture is a common form of field burning in tropical and forested areas.

The agriculture sector constituted approximately 36% of U.S. anthropogenic methane emissions in 2012.[47] From 1990 to 2012, methane emissions from agricultural sources increased by nearly 14% (see **Figure 2**). Enteric fermentation is the leading source of agricultural methane emissions, as well as the leading source of methane emissions from all industry sectors. Livestock manure management is the second-leading agricultural source (see **Figure 3**).

While best practices exist to reduce methane emitted from enteric fermentation (e.g., diet modification), it has been economically and technically challenging to systematically capture a significant portion of the methane emitted at this stage. There are, however, opportunities to reduce methane emissions from other agricultural sources, and efforts have focused on the second-largest agricultural source, manure management.

Anaerobic digestion (AD) systems[48] employed on stockpiles of manure at animal feeding operations may offer the most practical and economic method of capture. Operators have experience with AD systems partly because, for at least the last 20 years, USDA, DOE, and EPA have supported their use with financial and technical assistance (e.g., EPA/USDA's AgSTAR Program, established in 1994).[49] There are, however, some economic, operational, and safety concerns associated with the use of AD systems.[50]

The Obama Administration's Strategy takes a two-pronged approach to the reduction of agricultural methane from manure management. First, the Strategy supports a Biogas Roadmap, issued by USDA, EPA, and DOE on August 1, 2014, that outlines voluntary strategies to accelerate the adoption of AD systems and other technologies.[51] The Biogas Roadmap is a deliverable of an April 2013 Memorandum of Understanding between USDA and the Innovation Center for U.S. Dairy.[52] Second, the Strategy supports the continued use of previously established voluntary efforts (e.g., AD system deployment through assistance from numerous USDA programs).

The Strategy's goal is methane emission reduction, but the major agricultural source of methane emissions—enteric fermentation—is omitted from the Strategy. Some may wonder how much impact methane reduction from the agricultural sector can have if the major source is omitted. However, if the primary goal is cost-effective methane emission reduction, addressing manure management may be the most viable option for the agriculture sector at the moment.

Although federal support for AD systems using voluntary measures is not new, it is difficult to calculate the full impact of past and continued federal support. It is not clear that an adequate emissions baseline has been established among the appropriate federal entities for AD systems that receive federal support. An emissions baseline could allow for long-term analysis, which is necessary to gauge future impacts (e.g., number of AD systems, number of AD systems that are fully operational, amount of financial assistance provided, amount of methane captured, amount of methane flared, amount of methane used to generate electricity). Federal program data about AD systems tend to be disparate. The Strategy may give the federal government an opportunity to improve methods to document the impact of AD systems.

Beyond the availability and impact of adequate mitigation technologies for the agricultural sector, economic factors also may dampen the adoption of best practices. For these reasons, it could be argued that the establishment of a carbon market, the use of direct government payment programs for mitigating technologies like anaerobic digestion systems, and the development of voluntary mitigation-related contracts[53] could help alleviate costs and incentivize innovation. On the other hand, it may be that expansion of mitigation technologies such as anaerobic digestion systems face challenges larger than economics such as national infrastructure and cooperation with utilities or other industries that can use, but do not necessarily agree that they need, the product being sold.

Fossil Energy Sector[54]

Fossil energy sources of methane emissions include the following:

- **Petroleum Systems.** Methane emissions from petroleum systems are primarily associated with crude oil production, transportation, and refining operations. During each of these activities, methane is released to the atmosphere as fugitive emissions, vented emissions, emissions from operational accidents, and emissions from incomplete fuel combustion.
- **Natural Gas Systems.** The U.S. natural gas system encompasses hundreds of thousands of wells, hundreds of processing and liquefaction facilities, and over 1 million miles of transmission and distribution pipelines. Methane emissions arise from natural gas engine and turbine uncombusted exhaust, bleed and discharge emissions from pneumatic devices, and fugitive emissions from system components, as well as emissions from operational accidents.
- **Coal Mining.** Three types of coal mining-related activities release methane to the atmosphere: underground mining, surface mining, and post-mining (i.e., coal-handling) activities. While surface mines

account for the majority of U.S. coal production, underground coal mines contribute the largest share of methane emissions due to the higher methane concentrations in deeper coal seams.

The fossil energy sector constituted nearly 40% of U.S. anthropogenic methane emissions in 2012.[55] From 1990 through 2012, methane emissions from fossil energy sources have decreased by approximately 20% (see **Figure 2**). Natural gas systems are the leading source of emissions from the sector, and they have historically vied with enteric fermentation as the leading man-made source of methane emissions in the United States (see **Figure 3**). In its 2014 Inventory, EPA reported that methane emitted by the oil and gas sector had generally declined by 16% since 1990. However, EPA reports that it appears to be on the rise again, corresponding to increases in domestic onshore oil and gas production.[56]

Methane emissions from coal mining have remained relatively constant over the past several decades, accounting for approximately 10% of made-made emissions in the United States.

Taken together, the petroleum and natural gas industry (as shown in **Figure 4**) is one of the largest sources of methane emissions in the country, contributing in excess of 28% of U.S. anthropogenic methane emissions in 2012.[57]

Sources of emissions in the oil and gas sector include the following:

- **Upstream Production Sector.** Methane may be emitted while drilling through gas-bearing geologic formations, during drilling mud circulation, during well development (following well stimulation by hydraulic fracturing) when formation fluids and fracture fluids flow back to the surface, and from field treatment equipment that separates oil, gas, and water.
- **Midstream Processing and Transmission Sector.** Gathering lines connecting the wellhead to oil field treatment equipment that separates gas, oil, and water into product streams represent another source for fugitive methane and gas condensate emissions. Leaking valves, transmission lines, and pump stations add to this sector's emissions.
- **Downstream Distribution Sector.** Emissions from leaking distribution pipelines are most likely to occur from older pipelines. In 2012, there were more than 1.2 million miles of distribution mains in the United States. Of these, more than 32,000 miles of mains were

older cast iron or wrought iron, and more than 61,000 miles were unprotected steel.

Some companies in the oil and gas industries have made significant voluntary reductions in methane emissions over the past decade. By volume, some of the largest reductions have come using reduced emissions completions (or "green completions")[58] during hydraulic fracturing, leak detection and repair technologies at facilities and gas compressors, reduced venting of associated gas at oil wells, and the replacement of high-emitting pneumatic devices.

However, voluntary adoption of control techniques has been uneven across companies and regions. Consequently, in 2012, EPA promulgated emission standards for conventional pollutants (e.g., volatile organic compounds) for the oil and gas sector through a series of New Source Performance Standards (NSPS) and National Emissions Standards for Hazardous Air Pollutants.[59] These standards have the co-benefit of reducing methane emissions from certain new sources in some segments of the gas industry.[60]

Further, some states have established or proposed regulations that specifically address methane emissions from the oil and gas industry (e.g., Colorado, California, Ohio, and Wyoming, as well as a Western Governors' Association policy resolution).[61] Notwithstanding, many sources remain uncontrolled by state or federal standards.

BLM has issued rulemakings that address methane emissions on federal lands under the Mineral Leasing Act (MLA), but do not require practices to minimize methane emissions.[62] The MLA authorizes the Secretary of the Interior to lease onshore lands owned by the United States that contain fossil fuel deposits, with the federal government retaining title to the lands. The framework of the MLA provides BLM and the federal government with flexibility to use federal lands to help satisfy the nation's energy needs, while generating revenue for the federal government and protecting environmentally sensitive areas.

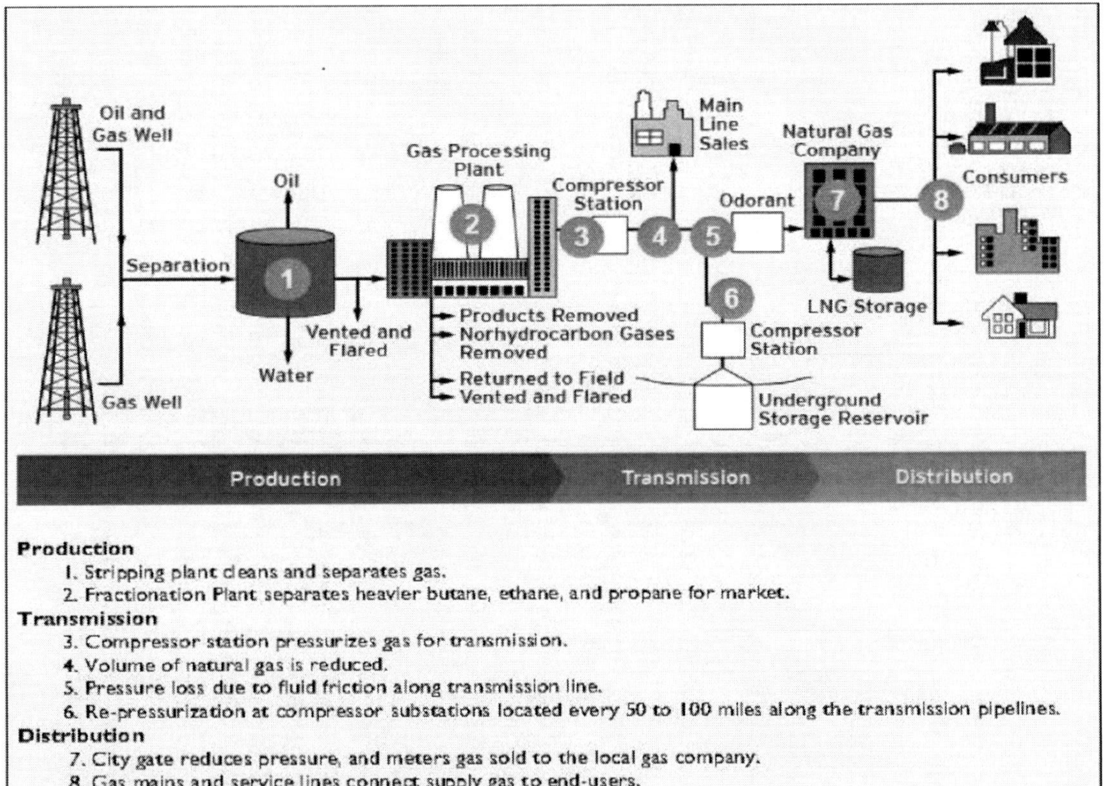

Source: DTE Energy, Natural Gas Processing, Delivery, and Storage.

Figure 4. Natural Gas Industry Sectors.

Existing BLM rulemakings affecting methane emissions include DOI, "Notice to Lessees and Operators of Onshore Federal and Indian Oil and Gas Leases (NTL-4A): Royalty or Compensation for Oil and Gas Loss."[63] This 1980 notice to operators of oil and gas leases outlines appropriate payment terms for losses of natural resources under the authority of the MLA. The notice lists circumstances wherein operators are authorized to vent or flare methane without incurring royalty obligations.

The Obama Administration's Strategy targets methane control in the fossil energy sector through a number of agencies. Since its release, the Administration has announced a series of steps it would take in 2015 and beyond. These include the following:[64]

- An EPA proposal to build on the 2012 NSPS "to set standards for methane and VOC emissions from new and modified oil and gas production sources, and natural gas processing and transmission sources"[65] (scheduled for release in the summer of 2015).
- An EPA proposal to extend VOC reduction requirements to existing oil and gas sources in ozone nonattainment areas and states in the Ozone Transport Region (scheduled for release in the summer of 2015). These requirements would be in the form of Control Techniques Guidelines, which states would be required to address in their State Implementation Plans.
- An EPA proposal to expand voluntary efforts under the Natural Gas STAR program and the Coalbed Methane Outreach Program.
- An EPA proposal to strengthen its Greenhouse Gas Reporting Program to require reporting in all segments of the industry (released on December 9, 2014).[66]
- A BLM proposal to update standards to reduce venting and flaring from oil and gas production on federal lands (scheduled for release in April 2015), and an Advance Notice of Proposed Rulemaking (ANPRM) to develop a program for the capture and sale, or disposal, of waste mine methane on lands leased by the federal government (released on April 28, 2014).[67]
- A Department of Transportation Pipeline and Hazardous Materials Safety Administration (PHMSA) proposal for natural gas pipeline safety standards (scheduled for release in 2015).
- The President's FY2016 budget request for $15 million in funding for DOE to develop and demonstrate more cost-effective technologies to detect and reduce losses from natural gas transmission and

distribution systems, and $10 million in funding to launch a program to enhance the quantification of emissions from natural gas infrastructure.

- DOE proposals to issue energy efficiency standards for natural gas and air compressors, advance research and development to bring down the cost of detecting leaks, work with Federal Energy Regulatory Commission to modernize natural gas infrastructure, and partner with local distribution companies to accelerate pipeline repair and replacement at the local level.
- DOE's Quadrennial Energy Review, which would include "additional policy recommendations and analysis on the environmental, safety, and economic benefits of investments that reduce natural gas system leakage."[68]

Many of these steps have yet to be proposed or are still in the very early stages of proposed rulemaking. Thus, many of the requirements have yet to be specified.

Waste Management Sector[69]

Waste management sources of methane emissions include the following:

- **Landfills.** Landfill gas—a mixture of roughly 50% methane and 50% CO_2, and including small amounts of other gases—is released into the atmosphere if not captured. The amount of gas produced at any given landfill depends on the amount of organic material in the waste, the landfill's design, the climate at the site of the landfill, and the operating practices used by the site's operator. In general, large amounts of organic waste and high levels of moisture in a landfill lead to greater gas production.
- **Wastewater Treatment.** Wastewater from domestic and industrial sources is commonly treated to remove soluble organic matter and other contaminants. Soluble organic matter is generally removed using biological processes in which microorganisms consume the organic matter for maintenance and growth. On occasion, these processes may be accidentally or deliberately managed under anaerobic conditions, producing methane.
- **Composting.** Composting of organic waste, such as food waste, garden (yard) and park waste, and sludge, is a common practice in the United States. Methane is formed in anaerobic sections of the

compost, but its impacts are generally mitigated due to oxygenation in the aerobic sections of the compost.

Waste management and treatment activities constituted approximately 21% of U.S. anthropogenic methane emissions in 2012. Landfills accounted for approximately 18% of total U.S. anthropogenic methane emissions in 2012, the third-largest contribution of any methane source in the United States. Their methane emissions totaled 102.8 million metric tons of CO_2 equivalents, 1.6% of total U.S. GHG emissions (see **Figure 1**).[70] Although substantial, methane emissions from landfills have declined 31% in recent years from a high of 149.3 million metric tons (MMT) in 1992 (see **Figure 3**). Additionally, wastewater treatment and composting of organic waste accounted for 2.2% and less than 1% of U.S. methane emissions, respectively.

Currently, landfill gas is captured at the nation's largest landfills. A common landfill gas capture system consists of an arrangement of vertical wells and horizontal collectors usually installed after a landfill cell has been capped. A 1996 CAA regulation known as the "Landfill Gas Rule" established New Source Performance Standards and Guidelines that require landfills with a 2.5- million-metric-ton design capacity that accepted waste after November 8, 1987, to capture and burn the gas.

The gas can either be flared or used for energy production—often it is used as fuel for electricity generation. In promulgating the 1996 rule, EPA said that the 2.5-million-metric-ton minimum "corresponds to cities greater than 100,000 people." The agency also stated that the regulations "will only affect less than 5 percent of all landfills," but would reduce emissions of methane by 37% at new landfills, and by 39% at existing facilities. Partly as a result of the 1996 regulation, and partly due to tax incentives and voluntary programs, there are 636 operational methane capture projects at landfills as of January 2014.[71] This represents roughly one-third of the 1,800 to 1,900 operational municipal solid waste landfills reported in operation by EPA.[72]

Whatever success existing regulations, tax incentives, and voluntary programs may be having, a significant amount of methane continues to be emitted even at landfills subject to the Landfill Gas Rule. In addition, there are few methane capture projects at smaller landfills and at landfills that ceased operation before November 1987 (those not covered under the CAA). The latter group, numbering in the tens of thousands of sites, poses a particular challenge.

Often, there is no responsible party who might implement a methane collection system if the site's original owner is no longer in business.

Thus, in response to the President's Climate Action Plan and the March 2014 methane Strategy document, EPA is in the process of reviewing the 1996 Landfill Gas Rule and Guideline. On June 30, 2014, the agency released a proposed revision to the NSPS for new and modified landfills and an ANPRM for existing landfills. EPA is under a consent decree to issue a final NSPS rule by March 30, 2015.

The NSPS would make no change in the universe of new or modified landfills subject to its requirements: the threshold would remain at 2.5 million metric tons of design capacity (or 2.5 million cubic meters of waste). The agency explains this decision by stating the following in the preamble to the proposed rule:

> [T]he cost burden for installing a collection and control system is more significant for small landfills, which are more often owned by small entities, than larger landfills. Certain costs to construct the gas collection system (e.g., flat fees for drill rig mobilization, and monitoring and construction costs) remain relatively constant regardless of the size of the landfill.
>
> For these reasons, the EPA is not proposing any changes to the current design capacity threshold of 2.5 million Mg [metric tons] and 2.5 million m^3 [cubic meters].[73]

But the proposed rule would require that a gas collection control system be installed and operational within 30 months after landfill gas emissions reach 40 metric tons of nonmethane organic compounds or more per year. Under the current NSPS, this threshold is 50 metric tons per year.

EPA expects few landfills to be affected by the proposed rule: according to an agency fact sheet, "EPA estimates that 17 new landfills would be subject to the proposed updated standards; however, only 11 would be required to install controls by 2023, based on their projected emissions."[74] The proposed standard would apply to a much smaller percentage of landfills than would the standard established in 2010 by the California Air Resources Board: that state standard, while structured differently, applies to any landfill with 450,000 or more tons of waste in place.

For existing landfills, EPA has not yet proposed revisions to the 1996 guideline. Rather, the agency has asked for "public input on methods to reduce emissions from existing municipal solid waste (MSW) landfills," and stated that it "intends to consider the information received in response to the ANPRM in evaluating whether additional changes beyond those in the proposed revisions for new sources are warranted."[75]

Issues in Measurement[76]

Unlike CO_2, where emissions are reported using well-tracked energy statistics,[77] methane is emitted to the atmosphere primarily through fugitive releases of the gas (e.g., leaks in infrastructure, vapors from landfills, eructation [i.e., belching or flatulence] from livestock). By definition, fugitive emissions are those which are diffuse, transitory, and elusive to capture. Thus, one of the greater difficulties in understanding the impacts of methane emissions is acquiring comprehensive and consistent observational data. Broadly, there are two approaches to measuring fugitive emissions of methane: "bottom-up" and "top-down."

Each approach has its respective strengths, weaknesses, and uncertainties. At present, the difference in data acquisition and analysis between these two approaches has returned competing—and occasionally conflicting— emission estimates.

- **Bottom-Up Approaches.** Bottom-up methodologies begin by directly measuring the emissions from a number of randomly selected pieces of equipment or activities to determine an average "emission factor" (i.e., formula) for each type. Emissions for the entire industry are then estimated by multiplying these emission factors by the activity levels for each component (e.g., the total population of livestock and its diet, the number of oil and gas wellheads and other components, or the volume of landfill material). Thus, while the inventory is supported by initial direct measurements, the final results are statistical averages derived through computation, and may not reflect actual emissions in the field. Because the quality of methane data for some sources can be either absent or highly variable, bottom-up emission estimates entail considerable uncertainty.
- **Top-Down Approaches.** Other studies use "top-down" methodologies for the calculation of leakage (e.g., satellite observations, ambient atmospheric measurements, and geostatistical inverse modeling). Atmospheric studies use data sets of ambient concentrations of methane and related hydrocarbons in the vicinity of the targeted industry, along with the known emission profiles for these gases from industry operations, to infer the emissions from the sectors. (That is, these methodologies capture methane emissions from all natural, agricultural, and industrial activities. Researchers must then parse data estimates for attribution to their appropriate

sources using such analyses as isotopic ratios or prevalence signatures from accompanying nonmethane hydrocarbons.) Due to the technology requirements, these studies are rarer than bottom-up approaches.

As with the bottom-up approaches, different top-down studies have returned different emission estimates. Further, reported emission rates have varied considerably across different regions, making source attribution highly uncertain at the national level.

In general, top-down methodologies have returned higher emission estimates than bottom-up approaches.

Reasons for this discrepancy include (1) researchers may be attributing naturally occurring methane emissions to man-made sources; (2) researchers may be attributing emissions inaccurately from one man-made sector to another; (3) atmospheric measurements may capture emissions that are not accounted for in EPA's Inventory (e.g., leakage from abandoned gas wells); (4) atmospheric measurements capture all the gross emitters, accidents, spills, and human errors, whereas component measurements use emission factors averaged over instances of "normal operation"; and (5) atmospheric studies may be biased to regions where there is known leakage.

Currently, the primary source of information on methane emissions in the United States is EPA's annually published *Inventory of U.S. Greenhouse Gas Emissions and Sinks.*[78]

EPA's Inventory is a "bottom-up" approach, employing commonly accepted emission factors and activity levels to calculate aggregate estimates for all source categories. Methodologies for the Inventory are based primarily on 2006 guidelines released by the IPCC[79] and supplemented with additional domestic information, where available.[80] Bottom-up methodologies are used also for EPA's Greenhouse Gas Reporting Program,[81] as well as the Energy Information Administration's (EIA's) *Natural Gas Annual.*[82]

Further, there are many examples of state,[83] local, and nongovernmental inventories[84] commissioned by a range of stakeholders, from regional and municipal agencies to community groups and academic institutions.

Due to the differences in emission factors, industry reporting, and levels of uncertainty, current inventories have returned a variety of emission estimates. These differences have also contributed to periodic revisions to EPA's Inventory, and these revisions have returned occasionally significant fluctuations in reporting (e.g., emission estimates in the Inventory for natural gas systems have fluctuated between 96.4 $MMTCO_2e$ and 221.2 $MMTCO_2e$

over the past five years due simply to changes in reporting methodology). Furthermore, EPA's Inventory has been challenged by a number of academic studies as under-reporting methane releases from man-made sources (as examples, a 2014 study by federal and academic researchers suggests that methane emissions from gas-producing areas in Colorado are as much as three times higher than EPA inventories;[85] a 2013 paper published by Harvard University researchers and federal scientists suggests that EPA's figures may be underestimated in some cases by as much as 50%;[86] and a February 2014 study by Stanford University researchers estimates that methane leakage from natural gas lines and other sources could be 50% higher than current EPA estimates).[87]

The White House Strategy proposes actions to enhance U.S. methane measurement in support of two broad goals: (1) improving the bottom-up emission data relevant for mitigation, and (2) advancing the science and technology for monitoring and validating atmospheric concentrations.[88] Actions in the Strategy include efforts to (1) enhance EPA's Inventory through new scientific evidence and data sources, (2) encourage the development of cost-effective measurement technologies through funding at DOE's Advanced Research Projects Agency— Energy, (3) maintain and further develop a nationwide methane monitoring network through funding at NOAA,[89] and (4) improve local, regional, and global emission modeling at EPA and DOE. EPA is already in the process of outlining a comprehensive strategy for significantly improving its methodology for estimating emissions from the oil and natural gas sector.

This effort is in response to recommendations made by an EPA Inspector General report.[90]

Moving forward, the Strategy will need to find a way to harmonize the differences in reporting between the bottom-up and top-down studies, dampen the artificial annual fluctuations in reported estimates, and provide more transparent and unbiased source data in order to guarantee credibility in EPA's Inventory for all stakeholders and fairness in any subsequent rulemaking.

CONCLUSION

For a variety of economic, environmental, and public health and safety reasons, various stakeholders have sought policies to reduce, capture, and reuse methane emissions for the past several decades.

But emissions of methane have proven to be difficult to measure and hard to control. Their naturally occurring presence in the environment, their wide and varied sources of emissions, and the fugitive nature of their release have contributed to these difficulties.

Nevertheless, methane is a valuable resource. Its dual nature as both pollutant and commodity has offered a unique opportunity for control, and many strategies have attempted to capitalize on the economics of recovery. Whether or not a given control strategy is effective and cost-efficient for a given industry has depended upon a number of factors including (1) the nature and extent of the emissions, (2) the technology available for capture, and (3) the market price for the recovered products (e.g., with declining natural gas prices, the economics of capture technology are less favorable). Some significant efforts have been made by industry and some state regulators to address methane emissions in their particular localities.

For its part, the federal government has contributed funding for research and technology development, voluntary guidelines and tax incentives for industry, rules for mineral rights lessees on federal lands, and, on occasion, air pollution standards.

While the most current data on domestic methane emissions show an 11% decrease over the past two decades, the source categories that have contributed to these reductions are few (i.e., landfills and natural gas production). While these industries have made noteworthy strides in emission reductions through a combination of best management practices and the co-benefits provided by other air pollution standards, they may represent only the "low-hanging fruit." Other sources of methane emissions may confront greater challenges. They may lack adequately demonstrated control technologies or cost-effective opportunities. They may not co-emit other air pollutants, and thus may lack the "co-benefits" of existing regulations. Some of these sources have seen recent or sustained increases in emissions (e.g., petroleum systems and manure management, respectively).

Other sources (e.g., enteric fermentation and wastewater treatment) have gone unaddressed for decades, as no economically viable technology solution has been offered.

The Obama Administration's recent Strategy—as well as a variety of recent proposals in Congress—attests to the continued interest in better emission assessments and appropriate policy responses. In considering strategies moving forward, it may be useful to ask the following questions:

1) Is the current set of methodologies used for measurement adequate enough to rationalize and/or prioritize the appropriate controls?
2) Is the projected rise in domestic fossil fuel production and petrochemical manufacturing significant enough to rationalize and/or prioritize additional controls?
3) Is the current rate of decline in observed emissions expected to continue, and, if so, is it sufficient enough to discharge the economic, environmental, and public health and safety concerns?
4) To what extent may recently promulgated and proposed rulemaking for air pollutants commonly co-emitted with methane also serve the co-benefit of reducing emissions of methane (e.g., the NSPS for VOCs on the oil and gas production sector and the petroleum refinery sector, and the revised National Ambient Air Quality Standard for ozone)?
5) If further reductions are under consideration for a given source category, should the response come from the federal government, state governments, the industries, or the market?
6) If further reductions are under consideration for a given source category, which policy tool(s) would be most appropriate: (1) increased funding for technology research, (2) expanded public-private demonstration projects with industry, (3) regionally targeted or state-sponsored guidance or rulemaking, (4) methane-specific state or federal command-and-control air pollution standards, or (5) economy-wide market-based mechanisms for either ozone or GHG controls? Do fluctuations in the market price of natural gas impact the choice of policy?
7) How should the burden of GHG reductions be distributed among the various GHG emissions sources, and how should methane's other environmental benefits (in comparison to oil and coal combustion) be weighed in this context?

Appendix A. Recent Legislative Proposals

Table A-1. A Selection of Recent Legislative Proposals with Methane Components

Cong.	Bill No.	Bill Title	Sponsor	Last Action	Methane Component
113	H.R. 3895 (H.R. 4286, H.R. 4304, and S. 2170 include similar provisions.)	Energy Exploration and Production to Achieve National Demand Act	Rep. Duncan, Jeff	2/12/2014: referred to House subcommittee.	The bill would have aimed to reduce or eliminate financial, regulatory, and technical barriers to energy exploration and production. It would have amended Section 302(g) of the CAA (42 U.S.C. 7602(g)) by adding "The term 'air pollutant' does not include carbon dioxide, water vapor, methane, nitrous oxide, hydrofluorocarbons, perfluorocarbons, or sulfur hexafluoride."
113	H.R. 3547	Consolidated Appropriations Act, 2014	Rep. Smith, Lamar	1/17/2014: became P.L. 113-76.	The bill prohibited any funds made available in the act to be used to promulgate or implement any regulation requiring the issuance of permits under title V of the CAA (42 U.S.C. 7661 et seq.) for carbon dioxide, nitrous oxide, water vapor, or methane emissions resulting from biological processes associated with livestock production.
113	H.R. 3424	Converting Methane Into Petroleum Act of 2013	Rep. Larson, John B.	10/30/2013: referred to House committee.	The bill would have amended the Internal Revenue Code to (1) include in the tax credit for investment in a qualifying gasification project any qualified methane conversion technology, and (2) allow an alternative fuel excise tax credit for liquid fuel produced through qualified methane conversion technology at a facility.

Cong.	Bill No.	Bill Title	Sponsor	Last Action	Methane Component
					It defined "qualified methane conversion technology" as a process for the molecular conversion of methane into other hydrocarbons and the use of such hydrocarbons to replace or reduce the quantity of petroleum present in motor vehicle fuel and for the production of chemicals.
113	H.Amdt. 507	Amendment to H.R. 2728	Rep. Holt, Rush	11/20/2013: House amendment not agreed to; failed by recorded vote: 190-230 (Roll no. 601).	Amendment would have allowed the Secretary of the Interior to issue regulations to reduce methane emissions from oil and gas drilling operations on public lands.
113	H.R. 1943	SUPER Act of 2013	Rep. Peters, Scott H.	5/10/2013: referred to House subcommittee.	The bill would have required the President to establish a Task Force on short-lived climate pollutants including methane. The Task Force would review existing and potential policies that promote emissions reduction, identify duplications and gaps in current programs, recommend efficiencies, and identify, compile, evaluate, and develop best practices.
113	H.Amdt. 512	Amendment to H.R. 1900	Rep. Tonko, Paul	11/21/2013: House amendment not agreed to; failed by recorded vote: 183-233 (Roll no. 605).	The amendment would have required an application for a natural gas pipeline to include information ensuring that methane emissions will be minimized before such application can be considered for approval.
113	H.R. 621	Ensuring Affordable Energy Act	Rep. Poe, Ted	2/15/2013: referred to House subcommittee.	The bill would have prohibited any funds appropriated or otherwise available for the Administrator of EPA from being used to implement or enforce (1) a cap-and-trade program, or (2) any statutory or regulatory

Table A-1. (Continued)

Cong.	Bill No.	Bill Title	Sponsor	Last Action	Methane Component
					requirement pertaining to emissions of one or more GHGs, including methane, from stationary sources.
113	H.R. 83	Consolidated and Further Continuing Appropriations Act, 2015	Rep. Christensen, Donna M.	12/16/2014: became P.L. 113-235.	The bill prohibited any funds made available in the act to be used to promulgate or implement any regulation requiring the issuance of permits under Title V of the CAA Act (42 U.S.C. 7661 et seq.) for carbon dioxide, nitrous oxide, water vapor, or methane emissions resulting from biological processes associated with livestock production and any provision in a rule requiring mandatory reporting of GHG emissions from manure management systems.
113	S. 2940	American Opportunity Carbon Fee Act	Sen. Whitehouse, Sheldon	11/19/2014: referred to Senate committee.	The bill would have amended the Internal Revenue Code to impose a fee on (1) fossil fuel products including coal, petroleum products, and natural gas, for carbon dioxide emissions; and (2) emissions of any greenhouse gas, including methane, from any greenhouse gas emission source. The bill would have established, implemented, and reported on a program to collect data on methane emissions by major nonnatural sources, including emissions attributable to the extraction and distribution of coal, petroleum products, and natural gas.

Cong.	Bill No.	Bill Title	Sponsor	Last Action	Methane Component
113	S. 2911	Super Pollutants Act of 2014	Sen. Murphy, Chris, and Sen. Collins, Susan	9/18/2014: referred to Senate committee.	The bill would have established a task force to review policies and measures to promote, and to develop best practices for, reduction of short-lived climate pollutants including methane.
113	S. 2739 (H.R. 860 includes similar provisions.)	Biogas Investment Tax Credit Act of 2014	Sen. Schumer, Charles	7/13/2014: referred to Senate committee.	The bill would have amended the Internal Revenue Code to allow for an energy tax credit through 2018 for investment in qualified biogas property, among other things. Eligible qualified biogas property was defined as including systems which use anaerobic digesters or other biological, chemical, thermal, or mechanical processes (alone or in combination) to convert biomass into methane for use as a fuel.
113	S. 805	Robert C. Byrd Mine and Workplace Safety and Health Act of 2013	Sen. Rockefeller, John D., IV	4/24/2013: referred to Senate committee.	The bill would have required the Secretary of Health and Human Services to promulgate regulations requiring that mining equipment used in a coal mine incorporate an atmospheric monitoring and recording device that samples and records the methane, oxygen, carbon monoxide and coal dust levels in the mine. The bill was introduced in the 112th Congress as S. 3443.
113	S. 332	Climate Protection Act of 2013	Sen. Sanders, Bernard	2/14/2013: referred to Senate committee.	The bill would have required the Administrator of EPA to impose a fee on any manufacturer, producer, or importer of a GHG polluting substance, and to submit to Congress a report describing the quantity of fugitive methane emissions emitted as a result of any leak in

Table A-1. (Continued)

Cong.	Bill No.	Bill Title	Sponsor	Last Action	Methane Component
					natural gas infrastructure, including recommendations for eliminating each such leak.
112	H.R. 6212	Biogas Investment Tax Credit Act of 2012	Rep. Kind, Ron	7/26/2012: referred to House committee.	The bill would have amended the Internal Revenue Code to allow for an energy tax credit through 2018 for investment in qualified biogas property. Eligible qualified biogas property was defined as including systems which use anaerobic digesters or other biological, chemical, thermal, or mechanical processes (alone or in combination) to convert biomass into methane for use as a fuel.
112	H.R. 2055	Consolidated Appropriations Act, 2012	Rep. Culberson, John Abney	12/23/2011: became P.L. 112-74.	The bill prohibited any funds made available in the act or any other act to be used to promulgate or implement any regulation requiring the issuance of permits under Title V of the CAA (42 U.S.C. 7661 et seq.) for carbon dioxide, nitrous oxide, water vapor, or methane emissions resulting from biological processes associated with livestock production.
112	H.R. 199 (S. 231 and S.Amdt. 215 to S. 493 include similar provisions.)	Protect America's Energy and Manufacturing Jobs Act of 2011	Rep. Capito, Shelley Moore	2/1/2011: referred to House subcommittee.	The bill would have suspended, during the two-year period beginning on the date of enactment of the act, any EPA action under the CAA with respect to carbon dioxide or methane pursuant to certain proceedings, other than with respect to motor vehicle emissions.

Cong.	Bill No.	Bill Title	Sponsor	Last Action	Methane Component
112	H.R. 153	Ensuring Affordable Energy Act	Rep. Poe, Ted	2/1/2011: referred to House subcommittee.	The bill would have prohibited any funds appropriated or otherwise available for the Administrator of the EPA from being used to implement or enforce (1) a cap-and-trade program, or (2) any statutory or regulatory requirement pertaining to emissions of one or more GHGs, including methane, from stationary sources.
112	H.R. 97 (H.R. 1023, H.R. 1287, H.R. 1292, H.R. 1777, H.R. 3400, H.R. 4301, S. 706, S. 1720, S. 2199, and S. 2365 include similar provisions.)	Free Industry Act	Rep. Blackburn, Marsha	2/1/2011: referred to House subcommittee.	The bill would have amended Section 302(g) of the CAA (42 U.S.C. 7602(g)) by adding "The term 'air pollutant' does not include carbon dioxide, water vapor, methane, nitrous oxide, hydrofluorocarbons, perfluorocarbons, or sulfur hexafluoride." Some similar bills focused solely on the exclusion of agricultural emissions.
112	S. 3443	Robert C. Byrd Mine and Workplace Safety and Health Act of 2012	Sen. Rockefeller, John D., IV	7/25/2012: referred to Senate committee.	The bill would have required the Secretary of Health and Human Services to promulgate regulations requiring that mining equipment used in a coal mine incorporate an atmospheric monitoring and recording device that samples and records the methane, oxygen, carbon monoxide and coal dust levels in the mine.
111	H.R. 6511	Ensuring Affordable Energy Act	Rep. Poe, Ted	12/9/2010: referred to House committee.	The bill would have prohibited any funds appropriated or otherwise available for the Administrator of EPA from being used to implement or enforce (1) a cap-and-trade program, or (2) any statutory or regulatory

Table A-1. (Continued)

Cong.	Bill No.	Bill Title	Sponsor	Last Action	Methane Component
					requirement pertaining to emissions of one or more GHGs, including methane, from stationary sources.
111	H.R. 4753 (S. 3072 includes similar provisions.)	Stationary Source Regulations Delay Act	Rep. Rahall, Nick J., II	3/4/2010: referred to House committee.	The bill would have suspended, during the two-year period beginning on the date of enactment of the act, any EPA action under the CAA with respect to carbon dioxide or methane pursuant to certain proceedings, other than with respect to motor vehicle emissions.
111	H.R. 3598	Energy and Water Research Integration Act	Rep. Gordon, Bart	12/1/2009: passed/agreed to in House by voice vote. 12/2/2009: referred to Senate committee.	The bill would have directed the Secretary of Energy to identify each of DOE's energy research, development, and demonstration programs and projects into which it would be appropriate to integrate water considerations. This included developing a Strategic Plan to evaluate and establish technical milestones for technologies to treat and utilize produced waters discharged from oil, natural gas, coal bed methane, and mining activities, among others.
111	H.R. 3534	Consolidated Land, Energy, and Aquatic Resources Act of 2010	Rep. Rahall, Nick J., II	7/30/2010: passed/agreed to in House by the Yeas and Nays: 209-193, 1 Present (Roll no. 513).	The bill, as introduced in the House, would have amended the Mineral Leasing Act (30 U.S.C. 201 et seq.) to require any federal coal lease and any modification of an existing coal lease to include terms that establish (1) the inclusion of methane released in conjunction

Cong.	Bill No.	Bill Title	Sponsor	Last Action	Methane Component
				8/4/2010: placed on Senate Legislative Calendar under General Orders.	with mining activities within the scope of the lease if the United States owns both the coal and gas resources, (2) a requirement that the lessee recover the associated methane to the maximum feasible extent, (3) a requirement to analyze the extent to which associated methane can be economically captured, and (4) that any federal coal mine methane resources that are captured and used or sold pursuant to a federal coal lease be subject to a royalty of not less than 12.5%. (These provisions were not included in the bill as reported or engrossed in the House or placed on the Senate calendar.)
111	H.R. 2454	American Clean Energy and Security Act of 2009	Rep. Waxman, Henry A.	6/26/2009: passed/agreed to in House; passed by recorded vote: 219-212 (Roll no. 477). 7/7/2009: placed on Senate Legislative Calendar under General Orders.	The bill would have set forth provisions concerning clean energy, energy efficiency, reducing global warming pollution, transitioning to a clean energy economy, and providing for agriculture and forestry related offsets. The bill would have required the Administrator of EPA to establish a cap-and-trade system for GHG emissions and set goals for reducing such emissions from covered sources by 83% of 2005 levels by 2050. Methane was defined as a GHG, given a GWP of 25, and included in the offset program. Any source category that was responsible for at least 10% of the uncapped methane emissions in 2005 was covered under the program.

Table A-1. (Continued)

Cong.	Bill No.	Bill Title	Sponsor	Last Action	Methane Component
					Methane recovered from landfill gas, wastewater treatment gas, coal mine methane used to generate electricity at or near the mine mouth, and qualified waste-to-energy projects were covered under the program's renewable electricity standard. The bill would have explicitly exempted agriculture from the cap-and-trade program.
111	H.R. 1426 (S. 527 includes similar provisions.)	To amend the Clean Air Act to prohibit the issuance of permits under title V of that Act for certain emissions from agricultural production	Rep. Lucas, Frank D.	3/12/2009: referred to House subcommittee.	The bill would have amended the CAA to prohibit the issuance of permits under Title V of that act for any carbon dioxide, nitrogen oxide, water vapor, or methane emissions resulting from biological processes associated with livestock production.
111	H.R. 1158	Biogas Production Incentive Act of 2009	Rep. Higgins, Brian	2/24/2009: referred to House committee.	The bill would have amended the Internal Revenue Code to allow for an energy tax credit for investment in qualified biogas property. Eligible qualified biogas property was defined as including systems that use anaerobic digesters or other biological, chemical, thermal, or mechanical processes (alone or in combination) to convert biomass into methane for use as a fuel.
111	H.R. 469	Produced Water Utilization Act of 2009	Rep. Hall, Ralph M.	2/11/2009: passed House on voice vote. 2/12/2009: received	The bill would have set forth provisions for the Secretary of Energy to encourage research, development, and demonstration of

Cong.	Bill No.	Bill Title	Sponsor	Last Action	Methane Component
				in Senate and referred to committee.	technologies to facilitate the utilization of water produced in connection with the development of domestic energy resources including coal bed methane, oil, natural gas, or any other substance to be used as an energy source.
111	H.R. 391	To amend the Clean Air Act to provide that greenhouse gases are not subject to the Act, and for other purposes	Rep. Blackburn, Marsha	1/14/2009: referred to House subcommittee.	The bill would have amended Section 302(g) of the CAA (42 U.S.C. 7602(g)) by adding "The term 'air pollutant' does not include carbon dioxide, water vapor, methane, nitrous oxide, hydrofluorocarbons, perfluorocarbons, or sulfur hexafluoride." Some similar bills focused on this definition solely with respect to agricultural emissions.
111	S. 2729	Clean Energy Partnerships Act of 2009	Sen. Stabenow, Debbie	11/4/2009: referred to Senate committee.	The bill would have set forth provisions to establish a program to govern the creation of credits from emission reductions from uncapped domestic sources and sinks. The bill would have required the Secretary of Agriculture and the Administrator of EPA to establish a cap-and-trade system for GHG emissions. Methane controls were an eligible offset activity, and included collection and combustion projects at mines, landfills, natural gas systems; manure management, composting, or anaerobic digestion; recycling and waste minimization; rice cultivation; and animal management practices including dietary modifications and pasture-based livestock systems.

Table A-1. (Continued)

Cong.	Bill No.	Bill Title	Sponsor	Last Action	Methane Component
					Further, the bill would have exempted the requirement to hold allowances for emissions resulting from the use of gas as an energy source if the gas was derived from a domestic methane offset project. The bill included research and demonstration assistance for approaches to reducing methane emissions associated with agricultural production (including livestock and crop production), including quantification of those reductions.
111	S. 1733	Clean Energy Jobs and American Power Act	Sen. Kerry, John F.	2/2/2010: reported out of the Committee on Environment and Public Works; placed on Senate Legislative Calendar under General Orders.	The bill would have set forth provisions concerning the reduction of global warming pollution, energy efficiency, renewable energy, water efficiency, green jobs and worker transition, and adaptation to the impacts of climate change. The bill would have required the Administrator of EPA to establish a cap-and-trade system for GHG emissions. Methane was defined as a GHG, given a GWP of 25, and included in the offset program. Eligible offset activity included methane collection and combustion projects at active underground coal mines, landfills, oil and natural gas systems, and manure management and biogas facilities.

Cong.	Bill No.	Bill Title	Sponsor	Last Action	Methane Component
111	S. 1462	American Clean Energy Leadership Act of 2009	Sen. Bingaman, Jeff	7/16/2009: placed on Senate Legislative Calendar under General Orders.	The bill would have required the Secretary of Energy, in consultation with other appropriate agencies, to support a civilian research program to develop advanced membrane technology that would be used in the separation of gases from applications, including those that pull gases from landfills and separate out methane.

Source: Congressional Research Service.

Notes: This section was prepared with the assistance of Lynn J. Cunningham, Information Research Specialist. The table lists only those bills that specifically mention "methane." Bills are ordered by Congress, split between the House and the Senate, and arranged by bill number starting with the most recent. If similar language is contained in different bills, the first bill introduced is presented in the table (with the subsequent bill numbers given in parentheses).

APPENDIX B. RECENT EXECUTIVE BRANCH INITIATIVES

A Selected Chronology of Recent Executive Branch Initiatives

June 25, 2013	White House released "The President's Climate Action Plan" (CAP) with a stated goal of "reducing methane emissions" through the development of an interagency strategy and the pursuit of collaborative approaches across the economy.
November 29, 2013	EPA released a Final Rule (FR) to amend the GHG Reporting Rule to raise the 100-year Global Warming Potential of methane from 21 to 25, in line with the 2007 IPCC AR4 findings agreed to by Parties to the UNFCCC. EPA, "2013 Revisions to the Greenhouse Gas Reporting Rule and Final Confidentiality Determinations for New or Substantially Revised Data Elements, FR," 78 Federal Register 71903.

Appendix B. (Continued)

March 28, 2014	White House released the "Strategy to Reduce Methane Emissions." The Strategy summarized the sources of methane emissions, committed to new steps to cut emissions, and outlined the Administration's efforts to improve the measurement of these emissions. The Strategy proposed steps to further cut methane emissions from landfills, coal mining, agriculture, and oil and gas systems through both voluntary actions and potential regulatory standards.
April 15, 2014	EPA released the Inventory of U.S. Greenhouse Gas Emissions and Sinks: 1990-2012, which reported that U.S. GHG emissions in 2012 totaled 6,526 million metric tons of carbon dioxide equivalents, of which 567.3 MMTCO2e, or about 9%, was methane.
April 15, 2014	EPA released for external peer review five technical white papers on potentially significant sources of methane emissions in the oil and gas sector (pneumatic devices, liquids unloading, well completions, compressors, and leak detection). The white papers focused on technical issues covering emissions and mitigation techniques that target methane and volatile organic compounds (VOCs).
April 28, 2014	BLM released an Advance Notice of Proposed Rulemaking (ANPRM) soliciting input on the development of a program to capture, sell, or otherwise dispose of coal-bed methane or methane gases that are released from coal or other type of mineral seam into the air during extraction operations. BLM, "Waste Mine Methane Capture, Use, Sale, or Destruction, ANPRM," 79 Federal Register 23923.
April 27, 2014	DOE hosted a roundtable under the CAP with representatives of labor and manufacturing organizations to discuss methane emissions from the midstream and downstream natural gas systems.
May 8, 2014	EPA proposed the "Gas STAR Gold" initiative, a program to certify oil and gas facilities that reduce emissions of methane.
May 20, 2014	DOE hosted a roundtable under the CAP with scientists and representatives from environmental groups and other nongovernmental organizations to discuss methane emissions from the natural gas sector.

July 17, 2014	EPA released a Proposed Rule (PR) that updated the standards of performance for new municipal solid waste landfills. The proposed limits for new landfills would require operators to capture two-thirds of their methane and air toxics emissions by 2023. EPA, "Standards of Performance for Municipal Solid Waste Landfills," 79 Federal Register 41795. EPA released an ANPRM soliciting input on methods to reduce methane and other emissions from existing municipal solid waste (MSW) landfills. EPA, "Emission Guidelines and Compliance Times for Municipal Solid Waste Landfills," Advance Notice of Proposed Rulemaking, 79 Federal Register 41772.
July 25, 2014	EPA's Office of Inspector General (OIG) released a report that stated that EPA "has placed little focus and attention on reducing methane emissions from pipelines in the natural gas distribution sector." EPA OIG, "Improvements Needed in EPA Efforts to Address Methane Emissions From Natural Gas Distribution Pipelines," Report No. 14-P-0324.
July 29, 2014	DOE announced a series of steps aimed at reducing methane emissions from natural gas transmission and distribution systems, including setting energy efficiency rules for new natural gas compressors and working with industry on research and development to improve natural gas system efficiency and reduce leaks. DOE, "Factsheet: An Initiative to Help Modernize Natural Gas Transmission and Distribution Infrastructure."
July 31, 2014	USDA released guidance for calculating GHG emissions from agriculture and forestry activities, part of its larger efforts to address agriculture's potential effects on climate change. USDA, "Quantifying Greenhouse Gas Fluxes in Agriculture and Forestry: Methods for Entity-Scale Inventory."
August 1, 2014	USDA, DOE, and EPA released the "Biogas Opportunities Roadmap: Voluntary Actions to Reduce Methane Emissions and Increase Energy Independence," a comprehensive list of programs, funding opportunities, and strategies to increase construction and use of methane-fed biogas reactors in the agriculture, wastewater treatment, landfill, and other sectors in part as a way to create a market for use of the gas.

Appendix B. (Continued)

December 9, 2014	EPA proposed amendments to subpart W of the Greenhouse Gas Reporting Program that would add reporting of GHG emissions from gathering and boosting systems, completions and workovers of oil wells using hydraulic fracturing, and blowdowns of natural gas transmission pipelines. EPA, "Greenhouse Gas Reporting Rule: 2015 Revisions and Confidentiality Determinations for Petroleum and Natural Gas Systems; Proposed Rule," 79 Federal Register 73148.
December 16, 2014	DOE's Advanced Research Projects Agency—Energy office announced $60 million in awarded grants for cutting-edge technology that will detect, locate, and measure methane emissions, among other initiatives.
January 14, 2015	EPA announced a series of steps the agency plans to take in 2015 to address methane emissions from the oil and gas sector, including (1) building on the 2012 NSPS for VOCs to address new and modified activities and equipment in the sector uncovered by the previous rule, (2) extending VOC reduction requirements to existing oil and gas sources in ozone nonattainment areas and states in the Ozone Transport Region (in the form of Control Techniques Guidelines, which states would need to address in their State Implementation Plans), and (3) expanding voluntary efforts under the Natural Gas STAR program.

Source: Congressional Research Service.

Note: Initiatives were selected based upon CRS's consideration of significance.

End Notes

[1] For a discussion of the sources of naturally occurring methane, see U.S. Environmental Protection Agency, *Methane and Nitrous Oxide Emissions from Natural Sources*, EPA 430-R-10-001, Washington, DC, April 2010.

[2] Natural gas extracted through drilling operations by the oil and gas industry is commonly composed of the following: methane, 70%-90%; ethane, propane, and butane, 0%-20%; carbon dioxide, 0%-8%; oxygen, 0%-0.2%; nitrogen, 0%- 5%; hydrogen sulfide, 0%-5%; and rare gases (e.g., A, He, Ne, Xe) in trace amounts. See the Natural Gas Supply Association's educational website, http://naturalgas.org/overview/background/, for further discussion of composition.

[3] The stated reduction values are estimates based on carbon dioxide emitted per unit of energy generated. See Energy Information Administration (EIA), Office of Oil and Gas. Carbon Monoxide: derived from EIA, *Emissions of Greenhouse Gases in the United States 2009.* Other pollutants derived from U.S. Environmental Protection Agency, *Compilation of Air Pollutant Emission Factors, Vol. 1, Stationary Point and Area Sources*, 1998.

[4] Hydraulic fracturing (hydrofracking, fracking, or fracing) is commonly defined as an oil or gas well completion process that directs pressurized fluids typically containing any combination of water, proppant, and any added chemicals to penetrate tight rock formations, such as shale or coal formations, in order to stimulate the oil or gas residing in the formation, and that subsequently requires high-rate, extended flowback to expel fracture fluids and solids. The National Petroleum Council estimates that hydraulic fracturing will account for nearly 70% of natural gas development within the next decade; see National Petroleum Council, *Prudent Development: Realizing the Potential of North America's Abundant Natural Gas and Oil Resources,* September 15, 2011. For more discussion on this technology, see the section on "Hydraulic Fracturing" in CRS Report R42333, *Marcellus Shale Gas: Development Potential and Water Management Issues and Laws*, by Mary Tiemann et al.

[5] These unconventional resources are commonly defined as follows: Tight sands gas is natural gas trapped in low-permeability and nonporous sandstones. Shale gas is natural gas trapped in shale deposits, a very fine-grained sedimentary rock that is easily breakable into thin, parallel layers. Coal-bed methane is natural gas trapped in coal seams. These resources are referred to as "unconventional" because, in the broadest sense, they are more difficult and/or less economical to extract than "conventional" natural gas, usually because the technology to reach them had not until recently been developed fully, or had been too expensive. For a more detailed discussion of these definitions, see the Natural Gas Supply Association's website, http://naturalgas.org/overview/unconventional-ng-resources/.

[6] Ernest J. Moniz et al., *The Future of Natural Gas: An Interdisciplinary MIT Study,* June 25, 2010.

[7] Support for natural gas production has come from the Obama White House. In his 2012 State of the Union speech, President Obama stated, "We have a supply of natural gas that can last America nearly 100 years, and my administration will take every possible action to safely develop this energy." President Barack Obama, "Remarks by the President in State of the Union Address," Washington, DC, January 24, 2012.

[8] U.S. Department of Labor, Occupational Safety and Health Administration (OSHA), *Chemical Sampling Information, Methane.*

[9] Health effects associated with exposure to ozone include premature death, heart failure, chronic respiratory damage, and premature aging of the lungs. Ozone may also exacerbate existing respiratory illnesses such as asthma and emphysema. See U.S. Environmental Protection

Agency, *Regulatory Impact Analysis: Final National Ambient Air Quality Standards for Ozone*, Research Triangle Park, NC, July 2011. While methane is a precursor to ground-level ozone formation, it is less reactive than other hydrocarbons. For further discussion on methane as an ozone precursor, see section "Methane: A Primer."

[10] As a greenhouse gas (GHG), methane emitted into the atmosphere absorbs terrestrial infrared radiation, which contributes to increased global warming and continuing climate change. For further discussion on methane as a GHG, see section "Methane: A Primer." For further discussion on climate change and its potential impacts, see CRS Report RL34266, *Climate Change: Science Highlights*, by Jane A. Leggett.

[11] Here, as elsewhere in the report, GHGs are quantified using a unit measurement called carbon dioxide equivalent (CO2e), wherein gases are indexed and aggregated against one unit of CO2. This indexing is referred to as the Global Warming Potential (GWP) of the gas. For more discussion on GWP, see section "Methane: A Primer."

[12] Like methane, volatile organic compounds (VOCs) are difficult to capture because of the diffuse nature of their releases. Also, leak prevention and recovery of VOCs may pay dividends in reducing product losses. Because the value of VOCs is highly variable, state and federal regulatory programs have required control of VOC emissions, even when the product value does not result in a net cost savings to the potential emitter (e.g., National Emission Standards for Hazardous Air Pollutants, vehicle standards, and State Implementation Plans for ozone precursor controls).

[13] See discussion under section "Fossil Energy Sector."

[14] U.S. Environmental Protection Agency, "Standards of Performance for New Stationary Sources and Guidelines for Control of Existing Sources: Municipal Solid Waste Landfills," 61 *Federal Register* 9905, March 12, 1996. The rule states that "the emissions of concern are non-methane organic compounds (NMOC) and methane," and that "methane emissions contribute to global climate change and can result in fires or explosions when they accumulate in structures on or off the landfill site."

[15] U.S. Department of the Interior, "Notice to Lessees and Operators of Onshore Federal and Indian Oil and Gas Leases (NTL-4A): Royalty or Compensation for Oil and Gas Loss," January 1, 1980.

[16] Clean Air Act, as amended, 42 U.S.C. 7401 et seq. For a summary of the CAA and EPA's air and radiation activities and its authorities, see EPA's website and CRS Report RL30853, *Clean Air Act: A Summary of the Act and Its Major Requirements*, by James E. McCarthy and Claudia Copeland.

[17] While methane is a precursor to ground-level ozone formation, it is less reactive than other hydrocarbons. Thus, EPA has officially excluded it from the definition of regulated hydrocarbons called volatile organic compounds (VOCs). See U.S. Environmental Protection Agency, *Conversion Factors for Hydrocarbon Emission Components*, Washington, DC, EPA-420-R-10-015, July 2010.

[18] *Massachusetts v. EPA*, 549 U.S. 497 (2007).

[19] U.S. Environmental Protection Agency, "Endangerment and Cause or Contribute Findings for Greenhouse Gases," 74 *Federal Register* 66496, December 15, 2009. The "endangerment" language in Sections 108, 111, 211, 213, 115, and 231 provides fundamental authorities. Also, Section 111(d) provides authority to control GHG emissions from existing sources, and Section 111(b) and (e) provide similar authorities for new sources.

[20] For example, U.S. Environmental Protection Agency and National Highway Traffic Safety Administration, "2017 and Later Model Year Light-Duty Vehicle Greenhouse Gas Emissions and Corporate Average Fuel Economy Standards; Final Rule," 77 *Federal*

Register 62623, October 15, 2012; and U.S. Environmental Protection Agency, "Carbon Pollution Emission Guidelines for Existing Stationary Sources: Electric Utility Generating Units; Proposed Rule," 79 *Federal Register* 34829, June 18, 2014.

[21] While the 1996 landfill standards are the only EPA rulemaking that specifically targets methane, many performance standards set on facilities for other air pollutants can have the co-benefit of reducing methane emissions. One example is EPA's August 2012 New Source Performance Standards (NSPS) for the "Crude Oil and Natural Gas Production" and the "Natural Gas Transmission and Storage" source categories. The NSPS regulate volatile organic compound (VOC) emissions from gas wells, compressors, and other equipment. See U.S. Environmental Protection Agency, "Oil and Natural Gas Sector: New Source Performance Standards and National Emission Standards for Hazardous Air Pollutants Reviews, Final Rule," 77 *Federal Register* 49489, August 16, 2012. For a summary of the NSPS, see CRS Report R42833, *Air Quality Issues in Natural Gas Systems*, by Richard K. Lattanzio.

[22] For example, the Family Farm Energy Conversion Act (S. 3714).

[23] For example, the Underground Coal Gasification and Unconventional Gas Research, Development and Demonstration Act (S. 2774) and the bill "to provide for the development and improvement of the recreation facilities and programs of Gateway National Recreation Area through the use of funds obtained from the development of methane gas resources within the Fountain Avenue Landfill site by the City of New York" (S. 2218) (P.L. 97-232).

[24] World Environment Policy Act of 1989 (S. 201).

[25] Clean Air Act Amendments of 1990 (S. 1630, P.L. 101-549). The findings were reported in U.S. Environmental Protection Agency, *Anthropogenic Methane Emissions in the United States: Estimates for 1990, Report to Congress*, EPA 430-R-93-003, 1993, which was expanded and replaced by U.S. Environmental Protection Agency, *U.S. Methane Emissions 1990-2020: Inventories, Projections, and Opportunities for Reductions*, EPA 430-R-99-013, 1999.

[26] CO2 Offsets Policy Enabling Act of 1990 (H.R. 5966).

[27] These programs are discussed in more detail in subsequent sections of this report.

[28] For a discussion of the performance of these and other voluntary programs, see the subsequent sections of this report on the respective industry sectors.

[29] Executive Office of the President, *The President's Climate Action Plan*, June 2013. For a summary of the CAP, see CRS Report R43120, *President Obama's Climate Action Plan*, coordinated by Jane A. Leggett.

[30] EOP, CAP, op cit., p. 10.

[31] Executive Office of the President, "Climate Action Plan: Strategy to Reduce Methane Emissions," March 2014.

[32] U.S. Department of Agriculture, "Fact Sheet: Biogas Opportunities Roadmap: Voluntary Actions to Reduce Methane Emissions, Increase Energy Independence and Grow the Economy," August 1, 2014.

[33] U.S. Environmental Protection Agency, "White Papers on Methane and VOC Emissions," April 15, 2014.

[34] Executive Office of the President, "FACT SHEET: Administration Takes Steps Forward on Climate Action Plan by Announcing Actions to Cut Methane Emissions," January 14, 2015.

[35] U.S. Department of Energy, "Factsheet: An Initiative to Help Modernize Natural Gas Transmission and Distribution Infrastructure," July 29, 2014.

[36] U.S. Department of the Interior, Bureau of Land Management, "Waste Mine Methane Capture, Use, Sale, or Destruction," 79 *Federal Register* 23923, April 28, 2014.

[37] U.S. Environmental Protection Agency, "Standards of Performance for Municipal Solid Waste Landfills," Proposed Rule, 79 *Federal Register* 41807, July 17, 2014; and U.S. Environmental Protection Agency, "Emission Guidelines and Compliance Times for Municipal Solid Waste Landfills," Advance Notice of Proposed Rulemaking, 79 *Federal Register* 41772, July 17, 2014.

[38] U.S. Environmental Protection Agency, *Conversion Factors for Hydrocarbon Emission Components*, Washington, DC, EPA-420-R-10-015, July 2010.

[39] Intergovernmental Panel on Climate Change (IPCC), *Climate Change 2013: The Physical Science Basis*, Working Group I Contribution to the Fifth Assessment Report of the Intergovernmental Panel on Climate Change.

[40] Intergovernmental Panel on Climate Change, *Climate Change 2007: The Physical Science Basis*, Working Group I Contribution to the Fourth Assessment Report of the Intergovernmental Panel on Climate Change.

[41] This description of GWP is summarized from James Bradbury et al., *Clearing the Air*, World Resources Institute, April 2013, p. 11.

[42] U.S. Environmental Protection Agency, "2013 Revisions to the Greenhouse Gas Reporting Rule and Final Confidentiality Determinations for New or Substantially Revised Data Elements," 78 *Federal Register* 71903, November 29, 2013.

[43] As calculated over 100 years. U.S. Environmental Protection Agency, *Inventory of U.S. Greenhouse Gas Emissions and Sinks: 1990-2012*, Washington, DC, EPA 430-R-14-003, April 15, 2014.

[44] For further discussion, see section "Issues in Measurement."

[45] As shown in Figure 3, there are many sources of methane emissions. For editorial reasons, this report focuses only on the most significant emitters. For greater discussion on smaller sources of emissions (such as forest fires, rice cultivation, stationary combustion, abandoned coal mines, petrochemical production, mobile combustion, and iron, steel and coke production), see EPA, Inventory, op cit.

[46] This section was authored by Kelsi Bracmort, Specialist in Agricultural Conservation and Natural Resources Policy.

[47] EPA, Inventory, op cit.

[48] An anaerobic digestion (AD) system feeds manure or other feedstock into a digester that breaks it down in a closed facility in the absence of oxygen to produce a variety of outputs including methane. The methane can then be captured for use as an energy source to produce heat or generate electricity. For more information on AD systems, see CRS Report R40667, *Anaerobic Digestion: Greenhouse Gas Emission Reduction and Energy Generation*, by Kelsi Bracmort.

[49] AgSTAR is a collaborative outreach effort of EPA, USDA, and DOE designed to reduce methane emissions from livestock waste management operations by promoting the use of biogas recovery systems. For more on the program, see information at http://www.epa.gov/agstar/. Federal funding opportunities available for AD systems are provided at http://www.epa.gov/agstar/tools/financing/index.html.

[50] AD system concerns include the expense associated with system construction and operation. Additionally, the technology requires daily operation and maintenance, some of which may exceed the technical capability of the average agricultural producer. Lastly, if the methane captured from an AD system is generated for electricity and sold to a utility, there may be utility collaboration concerns, especially regarding whether the utility will accept the electricity generated and at what price.

[51] U.S. Department of Agriculture, "Fact Sheet: Biogas Opportunities Roadmap: Voluntary Actions to Reduce Methane Emissions, Increase Energy Independence and Grow the Economy," August 1, 2014.

[52] U.S. Department of Agriculture, "USDA and Dairy Producers Renew Agreement to Reduce Greenhouse Gas Emissions and Increase Sustainability of Dairy Production," press release, April 24, 2013.

[53] ICF International, *Greenhouse Gas Mitigation Options and Costs for Agricultural Land and Animal Production within the United States*, February 2013. See chapter 3 of the report for more information on methane emission reduction potential of selected types of AD systems and break-even costs.

[54] This section was authored by Anthony Andrews, Specialist in Energy Policy, and Richard Lattanzio, Analyst in Environmental Policy.

[55] EPA, Inventory, op cit.

[56] EPA, Inventory, op cit.

[57] EPA, Inventory, op cit.

[58] A reduced emissions completion is "a well completion following fracturing or refracturing where gas flowback that is otherwise vented is captured, cleaned, and routed to the flow line or collection system, reinjected into the well or another well, used as an on-site fuel source, or used for other useful purpose that a purchased fuel or raw material would serve, with no direct release to the atmosphere." U.S. Environmental Protection Agency, "Oil and Natural Gas Sector: New Source Performance Standards and National Emission Standards for Hazardous Air Pollutants Reviews, Final Rule," 77 *Federal Register* 49489, August 16, 2012.

[59] EPA, NSPS, op cit.

[60] For further discussion, see CRS Report R42833, *Air Quality Issues in Natural Gas Systems*, by Richard K. Lattanzio.

[61] See Colorado's rules at http://www.colorado.gov/cs/Satellite/GovHickenlooper/CBON /1251648046456, California's rules at http://leginfo.legislature.ca.gov/faces/billNavClient. xhtml?bill_id=201320140SB1371, Ohio's proposed rules at http://www.epa.ohio.gov/dapc /genpermit/genpermits.aspx, Wyoming's proposed rules at http://deq.state proposedrules.asp, and the Western Governors' Association Policy Resolution 2015-02, Methane Emissions Regulation at http://westgov.org/images

[62] Mineral Leasing Act, as amended and supplemented, 30 U.S.C. 181 et seq. For a summary of the MLA and BLM's leasing activities, see BLM's website and CRS Report R40806, *Energy Projects on Federal Lands: Leasing and Authorization*, by Adam Vann.

[63] U.S. Department of the Interior, "Notice to Lessees and Operators of Onshore Federal and Indian Oil and Gas Leases (NTL-4A): Royalty or Compensation for Oil and Gas Loss," January 1, 1980.

[64] Executive Office of the President, "Fact Sheet: Administration Takes Steps Forward on Climate Action Plan by Announcing Actions to Cut Methane Emissions," January 14, 2015.

[65] EOP, Fact Sheet, op cit. For a discussion of the source categories under consideration, see U.S. Environmental Protection Agency, "White Papers on Methane and VOC Emissions," April 15, 2014, http://www.epa.gov/airquality/ oilandgas/whitepapers.html.

[66] U.S. Environmental Protection Agency, "Greenhouse Gas Reporting Rule: 2015 Revisions and Confidentiality Determinations for Petroleum and Natural Gas Systems; Proposed Rule," 79 *Federal Register* 73148, December 9, 2014.

[67] U.S. Department of the Interior, Bureau of Land Management, "Waste Mine Methane Capture, Use, Sale, or Destruction," 79 *Federal Register* 23923, April 28, 2014.

[68] EOP, Fact Sheet, op cit. For more discussion, see U.S. Department of Energy, "Factsheet: An Initiative to Help Modernize Natural Gas Transmission and Distribution Infrastructure."

[69] This section was authored by James E. McCarthy, Specialist in Environmental Policy.

[70] EPA, Inventory, op cit.

[71] U.S. Environmental Protection Agency, Landfill Methane Outreach Program, Energy Projects and Candidate Landfills, http://www.epa.gov/lmop/projects-candidates

[72] Slightly different estimates of the number of operational MSW landfills were presented at various points in EPA's July 2014 Advance Notice of Proposed Rulemaking (ANPRM) for existing MSW landfills. See U.S. Environmental Protection Agency, "Emission Guidelines and Compliance Times for Municipal Solid Waste Landfills," Advance Notice of Proposed Rulemaking, 79 *Federal Register* 41778, July 17, 2014.

[73] U.S. Environmental Protection Agency, "Standards of Performance for Municipal Solid Waste Landfills," Proposed Rule, 79 *Federal Register* 41807, July 17, 2014.

[74] U.S. Environmental Protection Agency, "Proposed Updates to the New Source Performance Standards for Municipal Solid Waste Landfills," fact sheet.

[75] U.S. Environmental Protection Agency, "Emission Guidelines and Compliance Times for Municipal Solid Waste Landfills," Advance Notice of Proposed Rulemaking, 79 *Federal Register* 41772, July 17, 2014.

[76] This section was authored by Richard Lattanzio, Analyst in Environmental Policy.

[77] According to EPA's Inventory, over 94% of CO2 emissions in 2012 are attributed to fossil fuel combustion for energy use. Further, many other CO2 emissions arise from similar combustion processes in various industries.

[78] EPA, Inventory, op cit.

[79] Intergovernmental Panel on Climate Change, *2006 IPCC Guidelines for National Greenhouse Gas Inventories.*

[80] EPA has undertaken its own emissions studies and modeling practices for the various U.S. sectors, including the development of the EPA Cattle Enteric Fermentation Model (CEFM); and the Gas Research Institute and U.S. Environmental Protection Agency, *Methane Emissions from the Natural Gas Industry, Volumes 1-15,* GRI-94/0257 and EPA 600/R-96-080, June 1996. EPA also references a multitude of academic literature for its calculations (see respective references in the Inventory). Further to this, EPA annually takes comments on its Inventory methodology, and adopts revisions where appropriate.

[81] In response to the Consolidated Appropriations Act, 2008 (H.R. 2764; P.L. 110-161), EPA issued the Greenhouse Gas Reporting Rule (74 *Federal Register* 56260), which requires reporting of GHG data and other relevant information from large sources and suppliers in the United States. Sectors include petroleum and natural gas systems, industrial and municipal landfills, and industrial wastewater treatment facilities, but not agriculture or forestry sources. See EPA GHG Reporting Program website, http://www.epa.gov/ghgreporting/.

[82] U.S. Energy Information Administration, *Natural Gas Annual* (various years).

[83] See, for example, Texas Commission on Environmental Quality, *Barnett Shale Phase Two Special Inventory Data*, 2011, http://www.tceq.texas.gov/airquality/point-source-ei/psei.html; Colorado Department of Natural Resources press release, "State to undertake major study on oil and gas emissions," January 9, 2013, http://dnr.state PressReleases.aspx; and California Greenhouse Gas Emission Inventory, http://www.arb.ca.gov/cc/inventory/ inventory.htm.

[84] See, for example, the Environmental Defense Fund, which, in conjunction with several universities and environmental engineering firms announced on October 10, 2012, the

launch of a comprehensive study of methane emissions from natural gas infrastructure in an effort to accumulate new data. These studies replicate the "component measurement" methodologies of EPA's Inventory, using current conditions and measurement practices. The first sector study—production—was published in 2013 (David T. Allen et al., "Measurement of Methane Emissions at Natural Gas Production Sites in the United States," *Proceedings of the National Academy of Sciences of the United States of America*, September 16, 2013). For more information, see Environmental Defense Fund's *Methane Leakage Study*, http://www.edf.org/methaneleakage.

[85] Gabrielle Patron et al., "A New Look at Methane and Non-methane Hydrocarbon Emissions from Oil and Natural Gas Operations in the Colorado Denver-Julesburg Basin," *Journal of Geophysical Research: Atmospheres*, accepted for publication, 2014.

[86] Scott Miller et al., "Anthropogenic Emissions of Methane in the United States," *Proceedings of the National Academy of Sciences of the United States of America*, November 25, 2013.

[87] Adam Brandt, et al., "Methane Leaks from North American Natural Gas Systems," *Science*, 343:6172, pp. 733-735, February 14, 2014.

[88] EOP, *Strategy*, op cit., pp. 11-14.

[89] The President's FY2015 budget requests $8 million above current funding of $6.5 million for this program.

[90] U.S. Environmental Protection Agency, Office of the Inspector General, "EPA Needs to Improve Air Emissions Data for the Oil and Natural Gas Production Sector," Report No. 13-P-0161, February 20, 2013.

In: Methane
Editor: Andrea Malone
ISBN: 978-1-63482-526-9

Chapter 2

BIOGAS OPPORTUNITIES ROADMAP: VOLUNTARY ACTIONS TO REDUCE METHANE EMISSIONS, INCREASE ENERGY INDEPENDENCE AND GROW THE ECONOMY*

U.S. Department of Agriculture

EXECUTIVE SUMMARY

Methane is both a potent greenhouse gas and a valuable source of energy. In the Climate Action Plan, President Obama directed the Administration to develop a comprehensive, interagency strategy to reduce methane emissions. In March 2014, the White House released the *Climate Action Plan - Strategy to Reduce Methane Emissions*. As part of the Strategy, the U.S. Department of Agriculture (USDA), the U.S. Environmental Protection Agency (EPA) and the U.S. Department of Energy (DOE) committed to work with industry leaders to formulate a biogas roadmap.

This *Biogas Opportunities Roadmap* builds on progress made to date to identify voluntary actions that can be taken to reduce methane emissions through the use of biogas systems and outlines strategies to overcome barriers to a robust biogas industry in the United States. It supports the U.S. dairy

* This is an edited, reformatted and augmented version of a document issued by the U.S. Department of Agriculture, U.S. Environmental Protection Agency, and U.S. Department of Energy, August 2014.

industry's voluntary 2008 goal to reduce its greenhouse gas emissions by 25% by 2020. This goal was a driver behind a partnership forged between the Dairy Industry and USDA in December 2009 and renewed in May 2013. As part of the 2013 renewal, the dairy industry also requested that USDA create a voluntary biogas roadmap to support this goal.

Biogas is a proven source of energy used in the United States and around the world for decades. The United States currently has more than 2,000 sites producing biogas. The Roadmap found that with the proper support, more than 11,000 additional biogas systems could be deployed in the United States. If fully realized, these biogas systems could produce enough energy to power more than 3 million American homes and reduce methane emissions equivalent to 4 to 54 million metric tons of greenhouse gas emissions in 2030, the annual emissions of between 800,000 and 11 million passenger vehicles.

In order to realize these opportunities, the Roadmap identifies actions the Federal government will take to increase the use of biogas to meet our renewable energy goals, strengthen the economy, and reduce methane emissions exclusively through voluntary actions. These include:

- *Promoting Biogas Utilization through Existing Agency Programs*: USDA, DOE, and EPA will use their existing programs as a vehicle to enhance the utilization of biogas systems in the U.S by ensuring that existing criteria for technical and financial assistance considers the benefits of biogas systems, leveraging over $10 million in research funding to enhance the economic viability and benefits of biogas systems and co-products, and strengthening programs that support the use of biogas for clean energy, transportation fuel, renewable chemicals and biobased products.
- *Fostering Investment in Biogas Systems*: To help overcome financial barriers to the widespread investment in biogas systems, USDA will lead efforts to improve the collection and analysis of industry financial and technical data needed to track the performance of anaerobic digesters, evaluate current loan and grant programs for opportunities to broaden the financing options available for biogas systems, and review Federal procurement guidelines to ensure that products of biogas systems are eligible for and promoted by applicable government procurement programs.
- *Strengthening Markets for Biogas Systems and System Products*: To strengthen U.S. markets for renewable energy and value-added non-energy products from biogas systems, USDA, DOE, and EPA will

review opportunities to overcome barriers to integrating biogas into electricity and renewable natural gas markets, for example, though modernizing existing Federal incentives provided for renewable energy generation. USDA, EPA, and DOE will also drive the creation of tools to help industry broaden the market development for energy and non-energy biogas systems products.

- *Improving Communication and Coordination*: In order to implement the strategies laid out in this document and promote strong coordination and messaging across Federal agencies, USDA will establish a Biogas Opportunities Roadmap Working Group that will include participation from DOE and EPA, as well as the dairy and biogas industries. The Working Group will collaborate with industry to publish a progress report in August 2015, which identifies and prioritizes policies and technology opportunities to expand the biogas industry and reduce greenhouse gas emissions.

The emissions intensity of the production of meat and milk in the United States is already much lower today than it was even a few decades ago[1]. Due to improvements in production efficiency, it's amongst the lowest in the world[2]. Enhancing the deployment of cost-effective technology to utilize biogas can increase revenues and reduce emissions, providing another "win-win" for farmers, communities, and the nation.

I. BIOGAS AND BIOGAS SYSTEMS

Biogas is primarily a mixture of methane and carbon dioxide produced by the bacterial decomposition of organic materials in the absence of oxygen. Depending on the source of organic matter, biogas typically contains 50-70% methane, 30-40% carbon dioxide, and trace amounts of other constituents, such as hydrogen sulfide, hydrogen, nitrogen, and siloxanes.

Today, methane accounts for nearly 9% of domestic greenhouse gas emissions. Thirty six percent of these emissions come from the agricultural sector, equivalent to over 200 million tons of carbon pollution. While methane's lifetime in the atmosphere is much shorter than carbon dioxide, it is more efficient at trapping radiation. Pound for pound, the comparative impact of methane on climate change is over 20 times greater than carbon dioxide over a 100-year period. Although U.S. methane emissions have decreased by

11 percent since 1990, they are projected to increase through 2030 if additional action is not taken.

Biogas systems have the potential to capture methane that would escape into the atmosphere and utilize it to create energy (e.g., electricity, heat, vehicle fuel). Other byproducts of biogas systems include non-energy products such as nutrient rich soil amendments, pelletized and pumpable fertilizers, and even feedstock for plastics and chemicals. Successful biogas systems capture and use gas from landfills and/or the anaerobic digestion of wastewater biosolids, animal manure, and other organics for energy. Each system includes both the infrastructure to manage the organic wastes as well as the equipment to generate energy from the resulting biogas. These systems have been used on a commercial scale in the United States since the late 1970s, when concerns over energy prices and U.S. dependence on oil spurred interest in the use of recovered biogas as a source of energy.

While the landfill gas energy industry has matured over the last 40 years due to third-party private investment, a strong project development community, and federal incentives, the biogas industry as a whole has not advanced at the same rate. There are currently more than 630 landfill gas energy projects in place across the United States, with more than 2,000 MW of installed capacity for electricity generation and more than 310 million cubic feet per day of gas delivered for industrial purposes.[3] Meanwhile, only 239 manure-based digesters are in operation across the United States. The landfill gas energy sector offers many lessons that could be applied to the biogas industry as a whole due to the similarities in project development and the technologies for processing and using the resultant biogas.

Differences between landfills and anaerobic digesters are clear; however, divisions based on feedstock sources are becoming blurred. While older biogas systems typically were designed to process one feedstock, new systems usually can accept a variety of organic materials. Traditional waste management systems and recycling or alternative processing options are now converging. Lines drawn between landfills, water resource recovery facilities, manure management, source-separated organics, and industrial waste streams are becoming harder to discern. Anaerobic digesters and biogas systems have become a hot topic for many local, state, and national discussions as policymakers recognize organic waste as a resource to use rather than a problem to manage.

Biogas Feedstocks

A growing awareness of the resource potential of organic material discarded each day in the U.S. is inspiring interest in using organic waste as biogas system feedstocks. Historically, the feedstocks for most biogas systems have been livestock manure, wastewater sludge and, in the case of landfills, municipal solid waste. While new projects continue to use these traditional feedstocks, many projects are also using source-separated and industrial organics as either a primary or supplemental feedstock. The primary biogas system feedstocks include:

- Livestock manure – dairy, swine, poultry, and beef.
- Municipal solid waste – mixed MSW delivered to landfill (~30% organics).
- Wastewater biosolids and primary sludge – by-product of water recovery treatment process.
- Food loss and waste – the amount of edible food, postharvest, that is available for human consumption but is not consumed for any reason.
- Food production residuals – by-products of the food production and processing industry.

Blending Feedstock for Biogas Systems: A Growing Trend in America

Figure 1. Harvest Energy Garden.

A growing number of existing and planned projects combine multiple feedstocks within a given biogas system. Some examples of this exciting new trend include:

- In Ithaca, New York, the local municipal water resource recovery facility is co-digesting food waste with wastewater biosolids to provide an alternate waste management option and boost biogas production.
- In Rutland, Massachusetts, a digester accepts manure from 300 cows as well as residuals from ice cream and salad dressing production to increase tipping fee revenue and biogas output.
- In Arlington, Texas, a biogas system blends landfill gas and wastewater digester gas to fuel an energy plant that powers the facility and supplies energy to the grid.
- In Janesville, Wisconsin, the local wastewater treatment facility digests biosolids and food wastes, using the biogas to both power microturbines for electricity production and a unit to produce vehicle fuel.
- At the Harvest Energy Garden in Lake Buena Vista, Florida, wastewater biosolids, fats, oils, grease, and food waste from Walt Disney World and surrounding communities are fed into the digester to recover energy and nutrients, manage odors, process biosolids, and produce a high-quality organic fertilizer.

As the biogas industry deploys more digester facilities across the country, the potential for blending feedstocks from various sources will increase due primarily to decreased hauling distances.

II. BENEFITS OF BIOGAS SYSTEMS

Biogas systems provide economic, energy, and environmental benefits for farms, businesses, and communities. These systems enable the capture and use of methane while also addressing waste management and nutrient recovery needs.

Biogas system products vary from energy (electricity, heat, fuel) to nutrient-rich soil amendments, pelletized and pumpable fertilizers, a renewable replacement for natural gas, and even feedstocks for renewable chemicals and bioplastics. The energy products typically come from the biogas, while the other products are made from the digested solid and liquid materials a biogas system produces.

Provide a Renewable Source of Energy

There are several different options for converting biogas to energy. Numerous factors such as project goals, local energy policies, infrastructure availability, and markets for renewable energy products will dictate what end use best fits the project. Unlike intermittent renewable energy alternatives such as wind and solar power, biogas delivers a continuous source of energy with a very high capacity factor. The flexibility and reliability of biogas systems are very important assets. Currently 37 states recognize biogas in their state renewable energy goals, and the U.S. government has set a target for 20 percent of the electricity consumed by Federal agencies to be from renewable energy by 2020. Biogas can assist in achieving these goals and provide many energy benefits. Specific commercially proven energy uses for biogas include:

- *Thermal applications:* Biogas is used directly on-site to heat digesters and buildings/maintenance shops, to fuel boilers or kilns, and to generate heat or steam.
- *Power generation:* Electricity is produced through an internal combustion engine, gas turbine, or microturbine technologies for on-site use or sale to the electric grid. Combined heat and power (CHP) systems increase overall energy efficiency of electricity systems by producing heat and electricity at the same time, which can be used for heating, cooling, dehumidification or other process applications. Unlike intermittent renewable energy sources, biogas systems are providing continuous dispatchable electricity onto the grid.
- *Industrial applications:* Biogas can be used in industrial applications to offset use of natural gas, propane, fuel oil, or other fossil fuels. Many industries such as sugar refineries, distilleries, dairies, and paper mills generate processing and waste water that can be digested directly on site. The resulting biogas can then be used for fuel in equipment such as boilers, kilns (e.g., cement, pottery, brick), sludge dryers, infrared heaters, paint shop oven burners, tunnel furnaces, process heaters, and blacksmithing forges or for other direct thermal applications.

- *Biomethane injection:* Upgraded and refined biogas, also called renewable natural gas (RNG), can be injected into existing natural gas networks.
- *Vehicle fuels:* Upgraded biogas can be converted to various vehicle fuels including compressed natural gas, liquefied natural gas, hydrogen, and liquid transportation fuels.

Figure 2. The Los Angeles County Sanitation District (LACSD) in California has operated a 50 megawatt (MW) landfill gas energy project at its Puente Hills Landfill since 1985, producing electricity for customers throughout Los Angeles.

Benefits of Combined Heat and Power

Vander Haak Dairy was the first Washington State dairy to install an anaerobic digester. The dairy utilizes on-farm waste and manure from two neighboring operations. Biogas generated is burned in a reciprocating engine. Thirty to sixty percent of the engine heat is used to heat the digester and the rest is used to dry bedding fiber and heat a house. Excess heat is available to meet additional needs of the dairy.

Biogas may also be used for fuel cells, which, with appropriate cleanup to remove trace contaminants, chemically convert biogas directly into electricity.

Like engine systems, fuel cell systems can be configured to produce heat as well as power. Certain fuel cell systems can also be configured to produce pure hydrogen, in addition to heat and power, known as trigeneration.

Currently in the United States, biogas fuels milk and recycling trucks, produces electricity for on-site and grid use, chills milk, heats greenhouses, produces steam, fires pottery and brick kilns, supplies pipeline quality gas, and provides fuel to local industrial plants. Projects range from small scale farm or community driven initiatives to multimillion-dollar private investments. Nearly 11,000 additional projects like these could be developed with the sources of biogas currently available in the United States.

Biogas Systems as 'BioRefineries'

There is a growing trend towards integrated biorefineries (biogas systems as sophisticated manufacturing centers) that are built to produce energy and high-value products as opposed to constructed as an add-on waste management process. These can involve a suite of technologies and processes to more efficiently and effectively process approved feedstocks to produce renewable fuels as well as marketable and valuable commodities and products, while potentially reducing environmental impacts. Primarily being developed by third-party private investors, these systems can be municipally owned, offering a good opportunity for public-private partnerships, or privately owned. As project developers look to more comprehensive solutions related to organic feedstocks, additional opportunities for biogas and co-product use are emerging. Some examples include:

- Biogas that is used to produce renewable hydrogen fuel for use in fuel cell applications.
- Biogas that is used as a feedstock for biodegradable plastics and intermediates for other bio-based product manufacturing.
- naerobic digester systems that enable algal biomass and advanced biofuel production. This could include biogas to generate electrical power to run algae production and biorefinery systems; excess heat offtake to stabilize and regulate water temperature systems for open raceway pond and photobioreactors; generator set exhaust that serves as the necessary CO_2 source for algae production; and recycled digester effluent that provides a needed nutrient source to promote algae biomass and lipid production.

Using Food Waste for Energy

USDA estimates that in 2010, approximately 133 billion pounds of food from U.S. retail food stores, restaurants, and homes went uneaten. This represents 31% of the 430 billion pounds of the available food supply at the retail and consumer levels in 2010, with retail-level losses accounting for 10% (43 billion pounds) and consumer-level losses for 21% (90 billion pounds) of the available food supply.

With the U.S. Food Waste Challenge, the U.S. Department of Agriculture (USDA) and the U.S. Environmental Protection Agency (EPA) have joined efforts to:

- Reduce food loss and waste,
- Recover wholesome food for human consumption, and
- Recycle food waste to other uses including animal feed, composting, and energy generation.

One objective of the U.S. Food Waste Challenge is to reduce the amount of food discarded to landfills. The EPA estimates that food waste is the single largest component of municipal solid waste going to landfills and that landfills are the third largest source of methane in the United States.

States, counties, and municipalities are helping to lead the way in reducing the amount of food waste discarded into landfills. Some are starting to mandate diversion of primarily commercial organics from landfills. Thus, more source-separated organics (SSO) are becoming available as feedstocks for biogas systems. Food production plants, universities, restaurants, hotels, and hospitals generate considerable volumes of organic wastes. Biogas systems can be designed and built specifically to process organic wastes on-site at these commercial facilities, or wastes from these sites can be transferred to serve as the feedstock for digesters at agricultural sites for improved food system resiliency. Some generators produce waste streams that are an economic liability to their operations but would be welcome financial additions to a biogas project, such as whey, residuals from bakery/brewery/winery, fats, oils and greases (FOG), due to the fact that these wastes produce high amounts of biogas.

The community digester at Sensenig Dairy in Kirkwood, Pennsylvania is fed six times a day with manure from cows, hogs and chickens, and community food waste.

The project has reduced emissions and operational costs, while creating additional revenue from the sale of carbon credits, fertilizer, and bedding.

Drive Economic Growth

Biogas systems offer a wide range of potential revenue streams, growing jobs and boosting economic development in the community. These systems can also improve rural infrastructure for waste management and distributed energy delivery improving community health, resiliency, and viability. Biogas systems can produce high-quality, concentrated liquid organic fertilizer for improved land management and increased crop yield, building and maintaining healthy and productive soils needed for sustainable food production. Along with generating revenues from the sale of renewable energy products, outputs from biogas systems can offer avoided costs of on-site electricity, heat, and transportation fuel. Renewable electricity can be sold into the power grid, and is often the primary driver for many biogas project investments. However, energy off-take contracts are often insufficient to fully finance a biogas system, and to be feasible many projects must realize the broader value of co-products, such as separated nutrients, marketable fertilizers and soil amendments. Separated fibers from the effluent stream can also reduce operational expenses or increase revenue through the production and sale of animal bedding. While niche markets exist for these products locally, developing more reliable national markets would reduce time for system payback, making project financing more attractive.

Additional opportunities exist for revenue generation from environmental attributes of the system, such as Renewable Energy Certificates (RECs) for electricity generation or Renewable Identification Number (RIN) credits under the Renewable Fuel Standard for the generation and use of biogas as vehicle fuel. There are also developing markets for carbon emission and nutrient offset credits, like that in California and the Chesapeake Bay watershed[4], which provide opportunities to offset regulatory compliance costs with voluntary installation of biogas systems.

All these potential financial returns can benefit project stakeholders and others involved in the biogas system. In addition, the federal government can provide environmental incentives to help defray infrastructure costs for systems which support conservation. For example, if a dairy farm installs a biogas system for enhanced manure management *and* renewable energy generation, cost-share funding from USDA's Environmental Quality Incentives Program (EQIP) could be combined with grant funding under USDA's Rural Energy for America Program (REAP) to help offset the investment.

Low-Emission Fuel for Vehicles

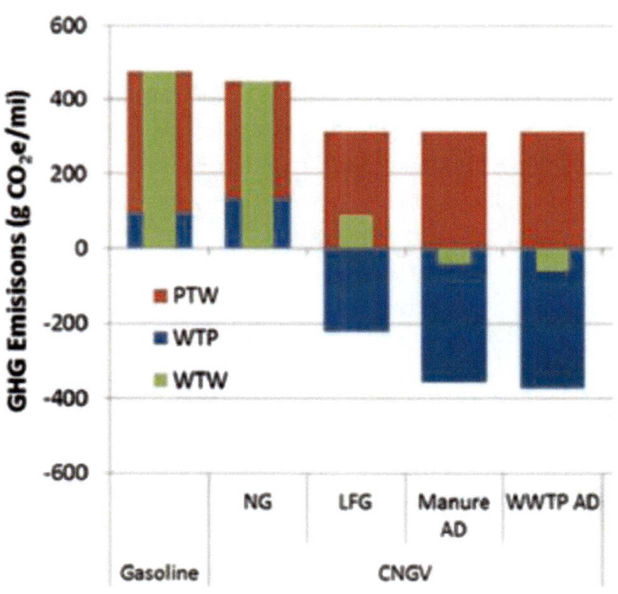

Figure 3.Well to Wheel Emissions, Biogas v. Fossil Fuel Sources (PTW- Pump to Wheels, WTP- Well to Pump, LFG-

On a well-to-wheel (WTW) basis, a truck fueled with fossil natural gas (NG) produces only slightly less CO_2 equivalent (CO_2e) emissions per mile traveled than one fueled with gasoline (Argonne National Lab, GREET 2013, http://GREET.es.anl.gov). If that same compressed natural gas vehicle (CNGV) were fueled with RNG produced from the anaerobic digestion of manure or at a wastewater treatment plant (WWTP), there would also be a significant reduction in CO_2e emissions from currently uncaptured and/or flared biogas, thereby resulting in negative CO_2e emissions.

Businesses and other organizations, such as universities and government facilities, can save on energy costs and achieve sustainability goals by choosing biogas as a direct fuel source in place of fossil fuels. Some end-users have saved millions of dollars over the duration of their biogas energy projects. Farmers and other companies who are recognized as leaders in sustainability and use of renewable energy may achieve indirect economic benefits through publicity of these accomplishments.

As with development of any energy project, biogas projects can benefit the local economy. Temporary jobs are created for the construction phase, while design and operation of the collection and energy recovery systems produce long-term jobs. Biogas energy projects involve engineers, construction firms, equipment vendors, and utilities or end users of the power produced. Some materials for the overall project may be purchased locally, and often local firms handle construction, electrical, plumbing, and other services.

Create Additional Revenue from Non-Energy Digester Products

In addition to energy, other potential revenue streams include nutrient recovery and management, tipping fees, thermal usage, bedding savings for farms, and carbon offsets, where available. Digestate liquids and solids (what remains after digestion) can produce additional economic benefits. The digestate has soil enhancement qualities and can be applied to growing crops, making it a marketable and valuable soil amendment. Reducing the need for synthetic fertilizers, the digestate delivers nutrients in a form that is more consistent, more readily absorbed, and more concentrated than raw manure. The use of digestate could provide a cost-savings to the farmer when compared to the purchase of synthetic fertilizers. Storage, mixing, pumping, and spreading digestate are easier than handling undigested organic materials, which can reduce energy demand and handling costs. Biogas production facilities designed to process landfill gas or source-separated organics (SSOs) provide economic benefits to the municipalities or waste management companies that own these facilities, as well as the broader community. Direct revenue sources include commercial tipping fees for SSOs.

An emerging benefit associated with biogas systems that use anaerobic digesters is the extraction of valuable nutrients, which supports environmentally and economically sound waste management. A number of systems, technologies, and procedures are available for nutrient recovery. The degree to which nutrients are removed depends on the value of the recovered nutrients, the need to produce clean water, and the economics of the technology used. Recovered nutrients offer an opportunity to create a "value-added" product that can be sold off-site as an organic amendment or as an organic fertilizer.

Cut Methane Emissions

Methane emissions in agricultural systems primarily come from three sources: livestock enteric fermentation, livestock manure waste, and rice cultivation. Manure management from dairy cattle, swine, and beef cattle operations in the United States accounts for 26% of all greenhouse gas emissions related to livestock sources.

Biogas systems can be used to capture methane that would escape into the atmosphere and contribute to climate change, and use it to create energy instead. The 239 livestock biogas systems currently operating in the U.S, reduce methane emissions by approximately 2 million metric tons of carbon dioxide equivalent annually. These projects provide enough renewable energy to power the equivalent of almost 70,000 average American homes.

The diversion of organics from landfills, collection of landfill gas and anaerobic digestion at waste water treatment plants can also decreases methane production and release.

Anaerobic digestion of livestock manure has been adopted by the State of California as an eligible project type for the generation of offsets under its statewide cap-and-trade program. This means that there is potentially a developing market demand for offsets from dairy and swine manure digester projects.

Protect the Environment

In addition to reducing methane emissions, some of the many environmental benefits of biogas systems include:

- Stabilization of nutrients for reduced water contamination risks, including substantial reduction of pathogens in manures and food wastes.
- Nutrient recovery and recycling.
- Reduction of odors during storage and decomposition.
- Providing a natural waste treatment process.
- Smaller physical footprint for organics waste processing versus composting.
- Reduced volume of waste for transport and land application.
- Efficient organic decomposition.

Digester systems protect America's waters by providing a step in broader biosolids treatment and nutrient management programs. The anaerobic digestion of manure and biosolids plays an important role in cost-effective wastewater treatment at thousands of facilities. Anaerobic digestion concentrates nutrients such as nitrogen and phosphorus, which can then be managed and diverted from water bodies to beneficial uses. With proper post-digestion nutrient management, biogas systems can thus improve water quality. Using an anaerobic digester to process organic wastes can help protect water quality. Pathogen levels can be reduced up to 99% compared to undigested manure. Anaerobic digestion is also an essential precursor to many advanced phosphorus and nitrogen separation technologies.

Biogas systems can offer significant improvements over traditional waste practices for organic material. While most organics other than manure find their way into landfills, companies and municipalities are looking to digesters for numerous environmental improvements.

Biogas systems must be properly designed and managed to operate effectively and avoid creating new environmental problems. Anaerobic digestion systems do not reduce the total amount of nutrients in the system or eliminate all organic wastes. For example, some of the organic nitrogen in the waste streams is converted to ammonium salts, which is easier for crops to utilize when incorporated into farmland. However, if not incorporated into the soil, ammonium salts can be converted to ammonia gas which is released, potentially at two to three times the rate of ammonia emissions from aerobic storage of organic waste streams. Similarly, some combustion equipment used to generate energy from biogas can increase air emissions as well. Therefore, it is important to evaluate the entire system holistically to determine impacts from pollutants of concern, including particulate matter, volatile organic compounds, nitrogen oxides (NO_x), ammonia, hydrogen sulfide, and greenhouse gases. In some instances the installation of a biogas system can necessitate additional controls and permitting requirements, which the project developers and stakeholders must be aware of and able to meet.

Overall, employing innovative digester systems with appropriate control of the nutrients, digestate solids and liquids, and air emissions could be a "win-win" for farmers, communities, the environment, and project investors. These efforts may lead to voluntary reductions in greenhouse gases and other air pollutants, pathogen load in runoff from farms, and the amount of organic wastes going to landfills. Communities that take appropriate actions to improve environmental quality, including the installation of biogas systems, can lower pollution control costs and extend the life of landfills. Economies of scale can be achieved by combining wastes from several sources.

Enhance Resilient Communities

Biogas systems can support sustainable communities by reducing methane emissions, improving water quality, producing a local source of renewable heat, electricity and fuel, and strengthening the local economy by reducing energy costs and generating revenue. They can also play a vital role in helping communities adapt and become more resilient to the effects of climate change. For example, the distributed nature of the biogas systems can increase the reliability of critical services – food, energy, waste management, wastewater treatment, and transportation – during and after disasters. Biogas systems are potentially less vulnerable to grid failures that can halt vital services. For example, a wastewater treatment or food production that is powered by onsite biogas could continue operation during a grid-wide power outage. Biogas system products could also be used to produce a renewable transportation fuel for routine use or should traditional sources be temporarily cut off.

Today's clean water agencies are increasingly considering how they can improve environmental performance, benefit their communities and improve their financial picture. The Water Resources Utility of the Future (UOTF) initiative encourages water utility leaders who are using innovative technologies and cutting-edge practices to focus on resource recovery including energy production, water reuse, green infrastructure or watershed-based approaches. The National Association of Clean Water Agencies, the Water Environment Research Foundation (WERF), and the Water Environment Federation (WEF) released the *Water Resources Utility of the Future ... Blueprint for Action* to define relevant issues, analyze key data, and offer recommendations for critical actions for the future. Currently more than 1,200 water resource recovery facilities have anaerobic digester systems, and more than 2,400 additional facilities could install an anaerobic digester on site. Working closely with the water resource utilities is one important way to help grow the biogas industry and enhance wastewater resiliency.[5]

Furthermore, putting food waste in digesters helps close the food system loop. Connecting food waste and nutrients back to the farm creates synergies and resiliency for agriculture's adaptation needs.

As an example, public and private partners in Columbus, Ohio added anaerobic digesters to the municipal wastewater treatment plant to process 300 tons per day of organic waste. The plant now produces enough biogas to generate one megawatt (MW) of electricity and 1,200 gasoline gallon equivalents (GGE) per day of CNG transportation fuel plus 90,000 gallons per day of nutrient-rich fertilizer for agricultural and landscape uses.

The Columbus, Ohio, Merchant Digester produces biogas from municipal wastewater biosolids, food and beverage wastes, and fats-oils-and greases. The biogas is converted to CNG and electricity to supply community needs for transportation, fuel, and power. The effluent is a high-value liquid fertilizer for use on farms and community landscapes.

III. Biogas Potential in the United States

There are vast organic resources available to feed biogas systems in the United States, with the primary feedstock sources being livestock manure, food waste, landfill gas, water resource recovery facility biosolids, and food production residuals. The decomposition of these organic materials can release methane, a potent greenhouse gas that has an effect on global temperatures that is over 20 times greater than carbon dioxide over a 100-year period. Already:

- Thirty-six percent of human-related methane emissions come from the agricultural sector in the United States, equivalent to more than 200 million tons of carbon dioxide pollution.[6]
- Municipal solid waste landfills account for approximately 100 million metric tons of carbon dioxide equivalent pollution.[7] While more than 600 landfills currently capture and use landfill gas for energy, hundreds of additional landfills are capturing their gas for compliance and safety but flare it without producing energy.
- In 2010, more than 130 billion pounds of food meant for human consumption at the retail and consumer levels was not consumed; this is the equivalent of approximately $160 billion worth of food.[8]
- In 2011, more than 34 million tons of food waste was landfilled or otherwise disposed of in ways that do not allow for nutrient recovery.
- More than 1,200 water resource recovery facilities across the United States use anaerobic digestion for biosolids management, thereby producing biogas that could be captured[9]
- Food production and processing facilities (e.g., milk processing, breweries, wineries, juice plants) produce large volumes of industrial organics as a by-product of their processes. While a number of these facilities have installed on-site digesters to manage these wastes, many more processors could produce biogas by installing digesters.

If captured and managed in a biogas system, these resources could yield substantial energy and bio-based product resources while providing environmentally sound management. According to U.S. Federal government and industry sources, the United States has more than 2,000 operational biogas systems out of more than 13,000 potential sites that could host a biogas system with manure, landfill gas and water recovery facility biosolids as feedstocks[x].

The potential for these systems to generate energy and reduce greenhouse gases is summarized in the following tables.

If the potential projects outlined in Figure 4 were fully realized, biogas could become a significant reliable renewable energy source. When taken together, these biogas sources could provide 41 billion kWh/year of electricity from 654 billion cubic feet of biogas/year. This is enough energy to power more than 3 million U.S. homes for one year or to produce the equivalent of 2.5 billion gallons of gasoline for vehicles.

Currently Operational and Potential Biogas Systems in the United States				
	Livestock Manure	Landfill Gas	Water Resource Recovery Facilities	Total
Currently Operational Biogas Systems	239[11]	636[12]	1,241[13]	2,116
Total Potential Number of Biogas Systems	8,241[14]	1,086[15]	3,681[16]	13,008

Figure 4. Currently Operational and Potential Biogas Systems in the United States Creating Energy.

Estimated Energy Potential from Biogas Sources in the United States[17]				
	Livestock Manure	Landfill Gas	Water Resource Recovery Facilities	Total
Biogas Production Potential (billion cubic feet/year)	257[18]	284[19]	113[20]	654
Annual Energy Production Potential (MMBTU/year)	142,000,000[21]	142,000,000[22]	67,000,000[23]	351,000,000
Annual Electricity Potential (billion kWh/year)	13.1[24]	22.5[25]	5.6[26]	41.2
Equivalent Residential Electricity Use (1000 homes/year) [27]	1,089	1,864	539	3,492
Potential Vehicle Fuel Gallons Displaced (million GGE)[28]	1,031	1,028	441	2,499

Figure 5. Energy Potential from Biogas Sources in the United States.

Cutting Carbon Pollution

Biogas capture from landfills, livestock operations and water resource recovery facilities can lead to significant reductions in U.S. greenhouse gas emissions. According to the *Global Mitigation of Non-CO2 Greenhouse Gases: 2010-2030*, annual methane reductions from the landfill, livestock and wastewater sectors could range from almost 4 to 54 million metric tons of greenhouse gas emissions in 2030, depending on the cost-effectiveness of various abatement options[29]. These reductions are equivalent to the annual greenhouse gas emissions of between 800,000 and 11 million passenger vehicles.

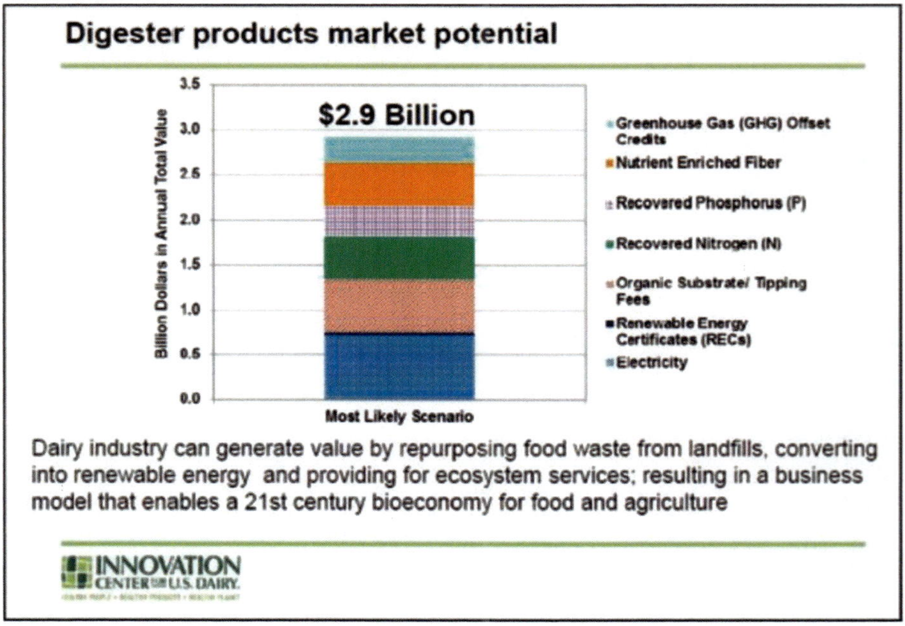

Figure 6. Dairy digester products market potential based on Informa Economics analysis.

Boosting the Economy

In estimating the market potential for full deployment, a lack of consolidated financial and technical data for the biogas industry limited analysis which could be done by federal agencies. Based on a survey with the industry and project developers reflecting current deployment, building those 11,000 potential systems would result in an estimated $33 billion in capital deployment for construction activity which would result in approximately 275,000 short-term construction jobs and 18,000 permanent jobs to build and run the digesters.[30] This number does not reflect the full market impact of biogas, which would also include energy and product sales and potential environmental credits. A complete economic analysis of the benefits of expanding biogas systems is not available; however, a study[31] examining the market potential from installing digesters on 2,647 dairy operations provides insight into the potential value.

Realizing the Potential of Biogas Systems

Fully realizing the market potential of biogas systems will take significant investment by livestock producers, municipalities, food producers, the private waste sector, and project developers. An integrated approach will be necessary to overcome the barriers limiting growth of the biogas industry. Critical efforts to promote development of biogas include:

- Support from federal agencies, including modifications or expansions of programs that advance biogas systems.
- Greater private investment in biogas systems.
- Development of broader markets for biogas and biogas system products.
- Increased emphasis on research and development to optimize systems.

IV. Primary Barriers to Realizing the Full Potential of a U.S. Biogas Industry

While there is a growing understanding among investors, policymakers, and the public of the value of investing in renewable energy systems, there remain significant barriers toward achieving a robust U.S. biogas industry.

– *Lack of Awareness of Biogas Benefits.* Investors, policy makers, and the public could benefit from gaining a deeper understanding of the value of investing in biogas systems and a biogas industry in the

United States. Greater public support for the adoption of biogas systems could result in more opportunity for biogas development.

- *Unpredictable Biogas Market Conditions.* Market unpredictability is a prime barrier toward greater investment in biogas systems. Unpredictability arises from multiple factors, including uncertainty and inconsistency in state and national energy policy, which restricts access to financial markets. Further market uncertainty arises when consumers perceive inconsistency in the quality or quantity of biogas systems or in the safety and quality of solid and liquid end products from digesters.
- *Lack of Market Maturity.* Underdeveloped markets for greenhouse gas reduction benefits pose a significant barrier toward biogas systems adoption. Additionally, immature markets for non-energy products– such as nutrient rich soil amendments, pelletized and pumpable fertilizers and feedstock for plastics – also reduce incentives to invest. Additionally,there is a need for a classification system for the use of digested solid and liquid residuals to encourage consumer confidence in product safety and consistency.
- *Lack of Full Valuation.* It is difficult for small generators to interconnect to the grid and to receive a fair market price that reflects the full environmental value provided. Overall, the high project costs without financial recognition of the non-energy servicescreate a barrier toward widespread investment in biogas systems.
- *Inconsistencies across Federal, State, and Local Governments.* Fragmentation of existing resources, regulatory authorities, and jurisdictions at the Federal, State, and local levels affect biogas system implementation. Additionally, fragmented efforts within and among Federal agencies to inform stakeholders, State and Tribal governments, and the public of the regulations, policies, practices, and potential funding of biogas systems create additional barriers.
- *Lack of Technical and Applied Research & Development.* The United States currently lacks adequate environmental, technical, and economic performance data related to biogas-system production of energy, co-products, greenhouse gas and other emissions, and water quality benefits. Consolidation of this data could help market analysis and underwriting. There is also a need for more advanced research in the United States related to renewable energy and biogas co-product benefits, including a better understanding of barriers by sector that

prevent full utilization of anaerobic digester capacity and digestion of feedstocks.

V. SOLUTIONS TO ENHANCE BIOGAS POTENTIAL

In order to help the private sector voluntarily realize the full potential of biogas systems, the Roadmap identifies near terms voluntary actions the government will take to promote biogas utilization through existing programs, foster investment and strengthen markets for biogas systems and products, and improve coordination and communication. Together, these actions will increase the use of biogas to meet our renewable energy goals, strengthen the economy, and reduce methane emissions.

Promote Biogas Utilization through Existing Agency Programs

A number of programs at USDA, DOE, and EPA are driving the development of biogas systems. AgSTAR and the Landfill Methane Outreach Program (LMOP) are dedicated to promoting biogas utilization from the livestock and landfill sectors in order to reduce greenhouse gas emissions. The Rural Energy for America Program, Environmental Quality Incentives Program, Bioenergy Program for Advanced Biofuel, Biorefinery Assistance Program, and Conservation Innovation Grants provide funding for biogas systems and components. USDA, DOE, and EPA will use existing programs as a vehicle to enhance the utilization of biogas systems in the U.S through:

Technical and Financial Assistance: USDA's Natural Resources Conservation Service (NRCS) provides technical and financial assistance to farmers and ranchers for voluntary conservation practices. The NRCS will conduct a full review of the standards used to determine which conservation practices are eligible to receive technical and financial assistance through the Environmental Quality Incentives Program (EQIP) and other programs to ensure that they recognize the full environmental benefits of modern anaerobic digesters. Accounting for these conservation benefits (e.g., methane destruction, manure separation and nutrient recovery, manure pipelines, and manure application) will enhance the amount of financial and technical assistance available to farmers and ranchers using biogas systems.

Research and New Technology: USDA's Agricultural Research Service and National Institute of Food and Agriculture will leverage over $10 million in research funding for anaerobic digesters to improve research for nutrient recovery, particularly nitrogen and phosphorus, from biodigester effluent and solids and investigate agronomic and economic viability of using captured nutrients as commercial fertilizers and soil amendments. USDA will also continue evaluating the carbon sequestration and soil productivity potential of biochar production from biodigester solids. These results will be communicated to stakeholders (e.g. industry, regulatory agencies, and private carbon market entities) to accelerate the adoption of anaerobic digester systems. In addition, DOE will further integrate biogas and biosolids systems into the Bioenergy Technologies Office program and develop a research plan to implement the recommendations of the Biomass Research & Development Technical Advisory Committee to accelerate development of bio-based products from biogas systems.

Promoting Biogas Systems in Federal Tools

- EPA recently released a National Agricultural Anaerobic Digestion Mapping Tool that allows users to view and analyze information about the current status and potential for biogas recovery systems in the agriculture sector.
- EPA is currently updating its Waste Reduction Model, which helps solid waste planners and organizations track and voluntarily report GHG emissions from several different waste management practices, to include anaerobic digestion.
- EPA is preparing updates to the Pacific Southwest region Co-Digestion Economic Analysis Tool, which assesses the initial validity of food waste co-digestion at wastewater treatment plants for the purposes of biogas production.

Partnerships: EPA will continue to engage stakeholders to address barriers to deploying biogas systems through existing programs, such as AgSTAR, the Landfill Methane Outreach Program, the Combined Heat and Power Partnership, and the Sustainable Materials Management program, including increasing outreach to state and regional partners on the benefits of biogas systems.

Transportation Fuel: DOE will include renewable natural gas from biogas as a clean energy option for research and development in the Vehicle Technology Office's Fuel and Lubricant Technologies Program to drive additional research on the utilization on biogas as a transportation fuel. DOE will also strengthen programs that support the use of renewable natural gas from biogas to compressed or liquid vehicle fuel directly; as feedstock to develop other renewable vehicle fuels (e.g., hydrogen, DME, etc.) and generate renewable liquid fuels (e.g., gasoline, diesel, jet fuel); and as a tool to increase fuel efficiency of vehicles. DOE will also increase the visibility of their existing commitment to support the use of renewable natural gas as a part of the Clean Cities Program's.

Renewable Energy: DOE will analyze the impact that biogas energy can have on electricity generation and fuel production in the U.S and its potential role as a drop-in biofuel and explore and map ways to integrate biogas with wind and solar for distributed renewable energy.

Biogas under the Renewable Fuel Standard

EPA has recognized the benefits of promoting net low-carbon fuels derived from biogas, and in a recent rulemaking EPA classified many sources of biogas as cellulosic feedstock for transportation fuels as part of the Renewable Fuel Standard (RFS). Cellulosic biofuels are the highest level of advanced biofuels specified in the RFS and achieve greater than 60% greenhouse gas (GHG) reductions as compared to the fossil fuels they replace. Use of biogas derived fuels in the transportation sector can substantially reduce GHG emissions and can serve to promote effective organic waste management, as well as efficient biogas production, recovery and utilization.

Further, use of biogas under the RFS can improve anaerobic digester economics by allowing biogas producers to generate Renewable Identification Numbers (RINs).

http://www.epa.gov/otaq/fuels/renewablefuels/r egulations.htm

Fostering Investment in Biogas Systems

High initial project costs create a barrier for the widespread investment in biogas systems. To begin to overcome this challenge, USDA, DOE, and EPA will take the following actions:

Propose NAICS Codes for Biogas Systems: The lack of NAICS codes for biogas systems has prevented the collection and analysis of industry financial and technical data needed to track the performance of anaerobic digesters. To address this, the Administration will assess the efficacy of developing NAICS codes for biogas systems, and if appropriate, submit a proposal for the development of a NAICS classification for biogas systems for consideration by the Economic Classification Policy Committee and the Office of Management and Budget.

Enhance Federal Financing: USDA and DOE will review applicable current loan and grant programs to enhance the financing options available for biogas systems. This includes exploring unique funding strategies for which biogas could qualify, including the Rural Utility Service's Energy Efficiency Conservation Loan Program along with the traditional Electric Loan Program, and improving access to capital under the Rural Energy for America Program. USDA will also work with the financial community through its partnership with the dairy industry to help them better understand the risks of biogas projects to encourage additional investment.

Lead by Example: To further the development and deployment of biogas systems, within 90 days, the USDA and partners will review federal procurement guidelines for alternative fuel use and renewable energy procurement and provide recommendations to CEQ and OMB, including the Office of Federal Procurement Policy, for ensuring that products of biogas systems are eligible for and promoted by applicable government procurement programs.

Strengthening Markets for Biogas Systems and Products

According to investors, market unpredictability is a prime barrier to greater investment in biogas systems. Immature markets for biogas energy and products are also limiting development of this technology. To strengthen U.S. markets for biogas systems energy and their value-added, non-energy products, such as recovered nutrients, fiber, and soil amendments, USDA, DOE, and EPA will take the following actions:

Accelerate the Use of Biogas in Clean Energy Markets: Already, 37 states consider biogas a renewable source of energy in their renewable energy targets. USDA, DOE, and EPA will continue to work with the appropriate state and local agencies to recognize biogas' role in supporting local and state environmental and renewable energy goals and ensure that biogas systems' contribution to greenhouse gas reductions, renewable energy generation, environmental improvements and energy security are recognized. USDA, DOE, and EPA will also review opportunities to overcome barriers to integrating biogas into electricity and renewable natural gas markets through the following mechanisms:

- Electric utility and natural gas interconnection standards;
- Interconnection fee structures;
- Natural gas pipeline injection standards;
- Fair market access and right to wheel provisions;
- Net-metering; and
- Current federal incentives provided for renewable energy generation.

Promote Products of Biogas Systems: USDA, EPA, and DOE will drive the creation of tools to broaden the market for non-energy biogas system products. These tools could include best management practices for digestate use and land application, particularly in targeted watersheds with nutrient trading potential.USDA, DOE, and EPA will also provide information on the ability of biogas system products to participate in markets that provide environmental benefits. This includes working to inform decisions that could increase the degree to which biogas receives credit related to renewable electricity, fuel, carbon reductions, and water quality improvements (e.g., RECs, RINs, carbon offsets, nutrient trading credits). USDA, DOE, and EPA will analyze markets for energy and non-energy products of biogas systems and the benefits these will generate. The energy and value-added products include:

- Electricity, heat, renewable natural gas, vehicle fuels.
- Nutrients, fertilizer, fiber, soil amendments.
- Liquid biofuels, renewable chemicals, intermediaries and bio-based products.

Improving Coordination & Communication

Strengthening communication across Federal agencies, state and local levels of government will be imperative to increase the adoption of biogas systems. To overcome this barrier, USDA, DOE, and EPA will:

Establish a Biogas Opportunities Roadmap Working Group: In order to implement the strategies laid out in this document and promote strong coordination and messaging across Federal agencies, USDA will establish a

Biogas Opportunities Roadmap Working Group that will include participation from DOE and EPA, as well as the dairy and biogas industry. The working group will commit to collaborating with industry to publish a progress report in August 2015 that identifies and prioritizes policies and technology opportunities to expand the biogas industry and reduce greenhouse gas emissions. A key component of this effort will be to assess existing and potential interagency cooperative structures, specifically EPA's and USDA's AgSTAR Program; DOE's and USDA's Biomass Research & Development Initiative; the EPA, USDA, and DOE "Biodigesters and Biogas" Workgroup; and the EPA, USDA, and USGS integrated nutrient management strategy.

Improve Information Sharing: USDA, DOE, and EPA will work together to improve existing information on biogas systems within government programs. This will include updating biogas data and links to resources that describe the benefits of biogas to reflect current knowledge and state of the industry on Federal websites. The Agencies will also provide guidance on incorporating biogas systems within existing technical assistance and market programs, including anaerobic digestion as a component of relevant project development tools. Agencies will also review current information related to renewable energy and other relevant initiatives to identify where additional coverage of biogas systems can help accelerate biogas system deployment.

Research and Development: USDA, DOE, and EPA will also continue to improve communication and coordination of research and development among government agencies, industry groups, and the public. Better communication of research results will aid industry's efforts to continue making advancements in the biogas sector. To initiate this process, the Biogas Opportunities Roadmap Working Group will identify research gaps in biogas and anaerobic digestion technology, including environmental benefits, market assessment, and performance standards. Examples for possible investigation could include:

- Nutrient capture technology and markets;
- U.S. biogas feedstock and biogas energy markets;
- Advanced biogas technology applications such as biochemical and algae production, carbon black, nano-fibers, biochar, fuel cell, and bio-plastic;
- Potential impact of biogas energy as first mover for other distributed renewable energy resources;
- Standards and testing for digester performance and solid and liquid residuals quality control;

- Biogas systems as infrastructure resiliency in municipal, natural disaster, and military applications;
- Logistics and infrastructure requirements for organic materials diverted from landfills to farms and community waste water treatment; and
- Biogas systems to improve rural and urban water resource recovery and treatment.

USDA, DOE, EPA Programs for Biogas Utilization

USDA, EPA, and DOE have targeted programs aimed at facilitating better communication and coordination. As the actions in the Roadmap are implemented, the agencies will utilize these programs to effectively disseminate new information to interested parties. For example, USDA and EPA will use the Food Waste Challenge to educate target audiences, especially organic waste generators, on the benefits of organics recycling using biogas systems. Additional examples of existing programs include:

USDA has programs from applied research to end use markets and financial and technical assistance programs to assist in deployment and assistance on biogas systems and, since 2009, has worked closely with the Dairy Industry to capture triple-bottom-line benefits with biogas systems.

USDA's primary programs for funding biogas systems are the Rural Energy for America Program, the Environmental Quality Incentives Program, Bioenergy Program for Advanced Biofuel, Biorefinery Assistance Program, and Conservation Innovation Grant, among others. Information on these programs, past investments and other tools for project development can be found at www.USDA.gov/Energy.

EPA currently provides a wide range of information related to biogas systems, including educational materials describing biogas systems and their benefits, profiles of biogas facilities, and technical information and tools to help stakeholders evaluate the feasibility of potential biogas projects. More information on these tools can be found at www.epa.gov/agstar.

DOE is developing advanced "drop-in" biofuels, which take advantage of existing infrastructure by providing nearly identical biobased substitutes for derived intermediates gasoline, diesel fuel, jet fuel, and chemicals and other products from crude oil. DOE has also made pioneering advances to reduce costs and establish best practices for harvesting, handling, and preprocessing a variety of crops for energy production. DOE's Bioenergy Technologies Office is focused on forming cost-share partnerships with key stakeholders to develop and demonstrate technologies for advanced biofuels production from lignocellulosic and algal biomass and waste streams. Additional information can be found at:

http://energy.gov/eere/transportation/bioenergy

http://www1.eere.energy.gov/financing/current_opportunities.html

CONCLUSION

Developing a viable biogas industry in the United States can boost the economy and provide a reliable, distributed source of renewable energy while reducing greenhouse gas emissions. Increasing production of biogas not only supports President Obama's Climate Action Plan goal of cutting methane emissions, but it also increases energy independence and security.

Biogas systems are currently installed primarily to manage wastes, but can also improve profitability for operations through energy and co-product sales, nutrient recovery and avoided energy costs. These new revenue streams come along with the added benefits of reducing greenhouse gas emissions, improving water quality, and limiting odors. Although 2,000 sites operate today, more than 11,000 additional biogas systems could be employed to handle organic waste and produce energy and biogas system co-products. Biogas can play a critical role in the sustainability and viability of communities throughout the U.S.

Realizing the full potential for the biogas industry will require support from federal agencies, greater investment, expanded markets for biogas and biogas products, and increased research and development. The benefits of biogas systems are clear. The task ahead is to reduce barriers and promote financial opportunities to move forward in developing a robust biogas industry.

End Notes

[1] Capper, J.L., Cady, R.A., Bauman, D.E. 2009. The environmental impact of dairy production: 1944 compared with 2007. Journal of animal science, 87 (6), 2160-2167.

[2] Gerber, P.J., Steinfeld, H., Henderson, B., Mottet, A., Opio, C., Dijkman, J., Falcucci, A., & Tempio, G. 2013. Tackling climate change through livestock – A global assessment of

emissions and mitigation opportunities. Food and Agriculture Organization of the United Nations (FAO), Rome.

[3] USEPA Landfill Methane Outreach Program.

[4] http://www.wri.org/publication/comparison-tables-state-nutrient-trading-programs-chesapeake-bay-watershed

[5] http://www.nacwa.org/index.php?option=com_content&view=article&id=1604&Itemid=250

[6] Climate Action Plan: Strategy to Reduce Methane Emissions, March 2014

[7] ibid.

[8] The Estimated Amount, Value, and Calories of Postharvest Food Losses at the Retail and Consumer Levels in the United States, USDA Economic Research Service, February 2014.

[9] Water Environment Federation.

[10] More biogas systems based on dedicated food production and processing facilities could also be built, although it is more difficult to estimate the full market potential due to a lack of industry data in those sectors.

[11] http://epa.gov/agstar/projects/index.html

[12] http://www.epa.gov/lmop/projects-candidates/index.html

[13] http://www.wrrfdata.org/biogas/biogasdata.php

[14] http://epa.gov/agstar/documents/biogas_recovery_systems_screenres.pdf

[15] http://www.epa.gov/lmop/projects-candidates/index.html (Added current + potential)

[16] Assuming facilities with an average flow > 1 million gallons per day are potential sites. http://www.wrrfdata.org/biogas/ and EPA 2008 Clean Watershed Needs Survey.

[17] The energy production potential and methane estimates were calculated using two difference mythologies. For further information on the methodology refer to the sources cited in each section.

[18] Assuming biogas from livestock manure is 60% methane; http://epa.gov/agstar/documents/biogas_recovery_systems_screenres.pdf

[19] Adding current + potential biogas production potential from http://www.epa.gov/lmop/projectscandidates/index.html (460 +317 mmscfd * 365 days/year = 283.605 billion ft3/year)

[20] Utilities of the Future Energy Findings, Steve Tarallo, et. al published by WERF under ENER6C13 estimate fall 2014.

[21] http://epa.gov/agstar/documents/biogas_recovery_systems_screenres.pdf

[22] Using potential biogas production above and a biogas energy content of 500Btu/scf. http://www.epa.gov/lmop/faq/landfill-gas.html

[23] Utilities of the Future Energy Findings, Steve Tarallo, et. al published by WERF under ENER6C13 estimate fall 2014.

[24] http://epa.gov/agstar/documents/biogas_recovery_systems_screenres.pdf

[25] Used current + potential electricity potential from: http://www.epa.gov/lmop/projects-candidates/index.html and converted MW to kWh/year using (MW*24 hrs/day * 365 days/year *.9 time online)

[26] http://www.werf.org/c/_FinalReportPDFs/OWSO/OWSO11C10.aspx

[27] Calculated based on 12,069 kWh/home/year (EIA2013a) - http://www.epa.gov/cleanenergy/energy- resources/refs.html

[28] Calculated using HHV of gasoline from: http://www.afdc.energy.gov/fuels/fuel_comparison_chart.pdf and a 90% system efficiency.

[29] Global Mitigation of Non-CO2 Greenhouse Gases: 2010-2030, EPA Report 430R13011, September 2013.

[30] based on industry survey of current project developers and assumes $3M capital cost, 25 construction job, 1.66 permanent job per project.

[31] National Market Value of Anaerobic Digester Products, Informa Economics, February 2013.

In: Methane
Editor: Andrea Malone
ISBN: 978-1-63482-526-9

Chapter 3

CLIMATE ACTION PLAN: STRATEGY TO REDUCE METHANE EMISSIONS*

Executive Office of the President

EXECUTIVE SUMMARY

Reducing methane emissions is a powerful way to take action on climate change; and putting methane to use can support local economies with a source of clean energy that generates revenue, spurs investment, improves safety, and leads to cleaner air. That is why in his Climate Action Plan, President Obama directed the Administration to develop a comprehensive, interagency strategy to cut methane emissions.

This document lays out that strategy – summarizing the sources of methane and trends in emissions; setting forth a plan to reduce both domestic and international methane emissions through incentive-based programs and the Administration's existing authorities; and, outlining the Administration's efforts to improve measurement of these emissions. This strategy also highlights examples of technologies and industry-led best practices that are already helping to cut methane emissions.

Today, methane accounts for nearly 9 percent of domestic greenhouse gas emissions. And although U.S. methane emissions have decreased by 11

* This is an edited, reformatted and augmented version of a document issued by the White House, March 2014.

percent since 1990, they are projected to increase through 2030 if additional action is not taken. As a key element of the Climate Action Plan, this strategy outlines new actions to reduce methane emissions. These actions will improve public health and safety while providing more energy to power our communities, farms, factories, and power plants. These steps will also make an important contribution to meeting the Administration goal of reducing U.S. greenhouse gas emissions in the range of 17 percent below 2005 levels by 2020. While the elements of the strategy will be further fleshed out in the coming months, Administration estimates show that steps along these lines could deliver greenhouse gas emissions reductions up to 90 million metric tons in 2020.

Through partnerships with industry, both at home and abroad, we have already demonstrated the technology and best practices to deliver substantial reductions in methane emissions. These cost-effective steps can deliver multiple benefits:

- *Economic Benefits:* Methane is the primary component of natural gas, so the recovery of methane that would otherwise be emitted can be productively used for power generation, heating or manufacturing. In addition, projects to reduce methane emissions can put people to work and spur investment in local economies.
- *Climate Change Benefits:* Every ton of methane in the atmosphere has a global warming effect that is more than 20 times greater than a ton of carbon dioxide. Thus, methane reductions yield important climate benefits, particularly in the near term.
- *Public Health:* Actions to reduce methane also improve the quality of the air we breathe. Methane is a contributor to ground level ozone, so cutting methane emissions reduces smog, which is associated with higher rates of asthma attacks. Moreover, methane is often co-emitted with volatile organic compounds, some of which are hazardous air pollutants, and many measures can cost-effectively reduce both pollutants.
- *Safety:* Improved safety and reduced methane emissions go hand-in-hand, as in our natural gas transmission and distributions systems and coal mining industries.

REDUCING METHANE EMISSIONS

The Administration is pursuing a targeted strategy that builds on progress to date and takes steps to further cut methane emissions from a number of key sources:

- *Landfills*: In the summer of 2014, the Environmental Protection Agency (EPA) will propose updated standards to reduce methane from new landfills and take public comment on whether to update standards for existing landfills. Through the Landfill Methane Outreach Program, EPA will further reduce methane emissions through voluntary programs – partnering with industry, state, and local leaders, many of whom are putting the methane waste to use powering their communities.
- *Coal Mines:* In April of 2014, the Interior Department's Bureau of Land Management (BLM) will release an Advanced Notice of Proposed Rulemaking (ANPRM) to gather public input on the development of a program for the capture and sale, or disposal, of waste mine methane on lands leased by the Federal government. In addition, EPA will continue to partner with industry through its voluntary program to reduce institutional, technical, regulatory, and financial barriers to beneficial methane recovery and use at coal mines.
- *Agriculture:* In June, in partnership with the dairy industry, the US Department of Agriculture (USDA), EPA and the Department of Energy (DOE) will jointly release a "Biogas Roadmap" outlining voluntary strategies to accelerate adoption of methane digesters and other cost-effective technologies to reduce U.S. dairy sector greenhouse gas emissions by 25 percent by 2020. USDA and EPA will also continue to support biodigester technology deployment by providing financial and technical assistance through voluntary programs.
- *Oil and Gas*: Building on the success of voluntary programs and targeted regulations in reducing methane emissions from the oil and gas sector, the Administration will take new actions to encourage additional cost-effective reductions. Key steps include:
    - In the spring of 2014, EPA will assess several potentially significant sources of methane and other emissions from the oil

and gas sector. EPA will solicit input from independent experts through a series of technical white papers, and in the fall of 2014, EPA will determine how best to pursue further methane reductions from these sources. If EPA decides to develop additional regulations, it will complete those regulations by the end of 2016. Through the Natural Gas STAR program, EPA will work with the industry to expand voluntary efforts to reduce methane emissions.

- o Later this year, the BLM will propose updated standards to reduce venting and flaring from oil and gas production on public lands.
- o As part of the Quadrennial Energy Review, and through DOE-convened roundtables, the Administration will identify "downstream" methane reduction opportunities.

Beyond these actions to reduce domestic methane emissions, the United States is also helping partners around the world to reduce methane emissions, including through the Climate and Clean Air Coalition and the Global Methane Initiative.

IMPROVING METHANE MEASUREMENT

Methane emissions come from diverse sources and sectors of the economy, unevenly dispersed across the landscape. These characteristics complicate measurement and attribution and lead to significant uncertainties in estimates of current and projected methane emissions. Better data collection and measurement will improve our understanding of methane sources and trends, and enable more effective management of opportunities to reduce methane emissions. Key steps under the strategy to improve data quality include:

- Developing new measurement technologies, including lower-cost emissions sensing equipment.
- Addressing areas of higher uncertainty in bottom-up inventories through additional data collection, direct emission measurements, and research and analysis.
- Enhancing top-down modeling and monitoring based on direct measurement of atmospheric concentrations.

I. Sources and Trends in Methane Emissions

Methane has a global warming potential more than 20 times greater than that of carbon dioxide, per metric ton; on this basis, emissions of methane from human-related sources were equivalent to approximately 560 million metric tons of carbon dioxide pollution in 2012 – making up nearly 9 percent of all the greenhouse gases emitted as a result of human activity in the United States1. Since 1990, methane emissions in the United States have decreased by 11 percent, even as many activities that can produce methane have increased. However, methane emissions are projected to increase to a level equivalent to over 620 million tons of carbon dioxide pollution in 2030 absent additional action to reduce emissions. The main sources of human-related methane emissions are agriculture (36 percent), natural gas systems (23 percent), landfills (18 percent), coal mining (10 percent), petroleum systems (6 percent), and wastewater treatment (2 percent) 2.

II. Reducing Methane Emissions

On June 25, 2013, the President issued a broad-based Climate Action Plan to cut the pollution that causes climate change and damages public health. The plan has three key pillars: cutting domestic greenhouse gas emissions, preparing the United States for the impacts of climate change, and continuing American leadership in international efforts to combat global climate change.

Low-cost technologies and best practices to recover methane and cut pollution are already widely available and used in key sectors. In addition to taking on climate change, reducing methane emissions has many other benefits. Recovered methane provides a local source of clean energy that can generate revenue and spur economic development. Reducing methane emissions can also improve safety and reduce local air pollution and odors. For example, landfill gas can be captured with collection systems and used for electricity generation or to provide fuel to a nearby factory. Methane recovered from coal mine degasification systems can be sold to natural gas pipelines or used on site for process heat. Methane generated from livestock manure can be captured through the use of biodigesters and used to generate electricity, avoiding fuel costs or providing a source of additional revenue. Methane that is vented or leaked from oil and natural gas production or processing facilities can be recovered through upgraded equipment and management practices that improve the company's bottom line.

This section outlines cost-effective opportunities to reduce methane emissions in four key sectors of our economy. It highlights new Administration actions to encourage voluntary emissions reductions and to set new standards where appropriate. The section also features key public-private sector initiatives that are cutting methane emissions or improving our understanding about specific sources of emissions. Finally, this section describes key U.S. efforts in helping international partners to reduce their methane emissions.

Reducing Emissions from Landfills

Municipal solid waste landfills are the third-largest source of human-related methane emissions in the United States, accounting for approximately 18 percent of methane emissions in 2012, equivalent to approximately 100 million metric tons of carbon dioxide pollution. Instead of allowing landfill gas to escape into the air, there is an opportunity to capture this gas and use it as a source of clean energy. In fact, standards and programs already in place have reduced landfill emission considerably, while creating jobs and improving public health. The Administration is committed to further reducing landfill emissions and tapping this important energy resource:

- *Updating Common Sense Rules to Reduce Landfill Emissions:* EPA will release a proposed update to its current standards for new municipal solid waste landfills in the summer of 2014, including assessing opportunities for further minimizing emissions when landfills are built or modified. Since there may be an even bigger opportunity for reducing methane emissions at existing landfills, EPA will also issue an Advanced Notice of Proposed Rulemaking (ANPRM) by June 2014 to engage industry and stakeholders on a range of approaches for cutting methane-rich landfill gases currently being emitted by existing facilities.
- *Enhancing Landfill Gas-to-Energy Projects:* EPA will continue to work with municipalities and landfill owners to advance cost-effective voluntary energy recovery projects at landfills through the Landfill Methane Outreach Program.
- *Reducing Landfill Waste:* Through the U.S. Food Waste Challenge, the USDA and EPA are challenging producers, processors, manufacturers, retailers, communities, and other government agencies to help reduce, recover, or recycle food waste. Less waste in our landfills, means less methane emissions, a win-win.

Case Study: Blue Ridge Renewable Energy Plant in Pennsylvania

In a true private-public partnership, landfill gas supplier IESI Blue Ridge Landfill, power purchaser Borough of Chambersburg and project developer PPL Renewable Energy (PPLRE) worked closely together to bring this 6.4-megawatt landfill gas electricity project online after only seven months of construction. In addition to designing, constructing, owning, and operating the LFG electricity plant at the landfill, PPLRE designed, permitted, and built the dedicated, 4-mile Express Generator Feeder from the plant to the Borough's Cree substation. Coming full circle, waste that Borough residents and businesses deposited in the landfill now supplies about 15 percent of its 11,000 customers' electric needs, plus the Borough was able to decrease the price of electricity those customers pay. In addition, the project generates 50,000 renewable energy credits annually toward meeting the state renewable energy goal.

Capturing Methane Emissions from Coal Mining

In 2012, 10 percent of human-related methane emissions came from the coal mining sector, equivalent to 56 million tons of carbon dioxide pollution. When recovered safely, coal mine methane can be a valuable, clean-burning source of energy. One of the most important co-benefits to reducing methane emissions at coal mines is increasing mine safety since uncontrolled methane emissions can cause fires and explosions. To further reduce coal mine methane emissions, the Department of Interior's BLM and the Environmental Protection Agency will take actions in two key areas:

- *Establishing a Program to Reduce Waste Coal Mine Emissions on Public Lands:* The BLM will release an Advanced Notice of Proposed Rulemaking (ANPRM) in April 2014, to solicit public input on the development of a program for the capture, sale, or disposal of waste mine methane from Federal coal leases and Federal leases for other solid minerals. The ANPRM will seek public input on preferred technology options for methane capture, whether the BLM should promote partnerships to capture or destroy waste mine methane, and how the agency could encourage cost-effective capture of methane from coal mines.
- *Overcoming Barriers to Reducing Coal Mining Emissions:* Since the Coalbed Methane Outreach Program's launch in 1994, the coal

mining industry has nearly doubled its total methane recovery and use. Currently, over 20 mines have installed methane degasification systems. The EPA will continue to work with industry through this voluntary program to encourage recovery and beneficial use of methane by helping to overcome institutional, technical, regulatory, and financial barriers. The EPA will also continue to coordinate with the Department of Labor's Mine Safety and Health Administration to ensure that implementation of methane recovery projects at coal mines is consistent with all applicable safety standards and with the BLM's efforts to facilitate coalbed methane capture and use on Federal lands.

Reducing Emissions from Agriculture

Thirty six percent of human-related methane emissions come from the agricultural sector in the United States, equivalent to over 200 million tons of carbon pollution. This strategy addresses emissions from agriculture exclusively through voluntary actions, not through regulations. The most important voluntary opportunities are through manure management with anaerobic digestion and biogas utilization. Biogas systems are proven and effective technology to process organic waste and generate renewable energy. They can reduce the risk of potential air and water quality issues while providing additional revenue for the operation. Yet, there are still relatively few digesters in operation on farms across America. To encourage adoption of this technology, the Administration is committed to promoting additional, cost-effective actions to reduce methane emissions through voluntary partnerships and programs, including:

- *Developing a Biogas Roadmap:* This June, in partnership with the dairy industry, USDA, EPA and DOE will jointly release a Biogas Roadmap outlining voluntary strategies to accelerate the adoption of biogas systems and other cost-effective technologies to reduce greenhouse gas emissions. This work will support the U.S. dairy industry goal to reduce greenhouse gas emissions by 25 percent by the year 2020 for the entire value chain.
- *Enhancing Biogas System Deployment:* USDA will continue to support biogas system deployment by providing financial and technical assistance through the Natural Resources Conservation Service's Environmental Quality Incentive Program and Rural Development's Rural Energy for America Program, Bioenergy

Program for Advanced Biofuels, and Biorefinery Assistance Program. The Natural Resources Conservation Service Conservation's Innovation Grants program is already catalyzing reductions in methane emissions through projects on rice cultivation and manure management. Through the AgSTAR program, EPA and USDA will continue working together to overcome barriers to expanding the use of agricultural biogas systems, and provide a forum for industry and other stakeholders to access information.

The Dairy Innovation Center
"Dairy Power" Project

The Innovation Center for U.S. Dairy's "Dairy PowerTM "project is focused on realizing the significant potential of anaerobic digester systems that can produce clean energy and value-added products, generate revenue for dairy producers and create jobs. The Dairy Power project report findings show a $3 billion market potential through the products and co-products developed by mature digester systems that process manure and commercial food waste, with additional value for potential nutrient trading markets, and renewable energy and low-carbon energy production incentives.

Reducing Methane Emissions from the Oil and Natural Gas Sectors

In 2012, 28 percent of methane emissions were attributed to the oil and natural gas sectors. Methane equivalent to 127 million tons of carbon dioxide pollution was emitted from production, processing, transmission, storage, and distribution of natural gas. Methane equivalent to 32 million tons of carbon dioxide pollution was emitted from production and refining, of oil. Within the natural gas industry, approximately 31 percent of this methane came from production sources, 15 percent from processing, 34 percent from the transmission and storage, and 20 percent from distribution. As our use of natural gas in manufacturing, transportation, and power generation increases – creating jobs, reducing costs, cutting carbon pollution, and reducing dependence on foreign oil in our nation – we must continue to build on progress in reducing methane emissions from this vital sector of our economy.

There are cost-effective technologies and best management practices to capture methane from venting and leaks across the entire oil and natural gas value chain. These range from equipment upgrades or replacements, to process or operational changes. Building on progress to date, the Administration will undertake new steps, including:

- *Working with States:* States are the primary regulators of many aspects of oil and gas production activities and the distribution of natural gas. DOE and EPA will continue to provide technical assistance in support of effective state policy actions to reduce emissions, and to encourage broader adoption of proven mitigation strategies.
- *Building on Common-Sense Federal Standards*: Since 2012, the EPA has taken a series of steps to address air pollution from the oil and gas sector. On April 17, 2012, the EPA issued final regulations to reduce the emissions of volatile organic compounds (VOCs), some of which are hazardous air pollutants, establishing, among other things, the first Federal air pollution standards for natural gas wells that are hydraulically fractured, along with requirements for other sources, such as compressors, that were not previously regulated at the Federal level. Although these regulations targeted VOCs, they also reduced methane emissions substantially. EPA estimates that when fully implemented in 2015, the 2012 rules will decrease methane equivalent to 33 million tons of carbon pollution per year. Going forward, the EPA will deploy a carefully selected combination of policy tools to maximize cost-effective methane and VOC reductions from the oil and gas sector.

  During the spring of 2014, the EPA will release a series of white papers on several potentially significant sources of methane in the oil and gas sector and solicit input from independent experts. The papers will focus on technical issues, covering emissions and control technologies that target both VOC and methane —with particular focus on oil and co-producing wells, liquids unloading, leaks, pneumatic devices and compressors. The agency will use these technical documents to solidify its understanding of these potentially significant sources of methane. This robust technical understanding will allow the agency to fully evaluate the range of policy mechanisms that will cost-effectively cut methane waste and emissions. The EPA will make peer reviewer comments available this summer. This fall, the EPA will determine what if any regulatory authorities, including setting standards under section 111 of the Clean Air Act or issuing Control Techniques Guidelines under section 182 of the Act, the agency will apply to emissions from these sources. If the agency determines to follow a regulatory course of action, it will undertake a schedule that will ensure that both rulemaking and any ensuing regulatory requirements for the states are completed by the end of 2016.

- Enhanced Partnerships and Stakeholder Engagement: The Administration will work collaboratively with key stakeholders to reduce methane emissions from natural gas systems.
  - In coordination with the Executive Office of the President and other Federal agencies, Secretary of Energy Ernest Moniz hosted a roundtable discussion on March 19, 2014 with leaders from industry, state governments, academic researchers, non-governmental organizations, and labor. DOE will sponsor additional roundtable discussions with stakeholders, with the primary objective of accelerating the adoption of best practices for reducing methane emissions from natural gas systems. Through these DOE roundtables the Administration also aims to:
    - Promote a common understanding of methane emissions from natural gas systems and related abatement opportunities.
    - Develop strategies for cost-effectively reducing methane emissions from processing, transmission and storage and distribution segments of the supply chain.
    - Catalyze greater action and engagement by policymakers at all levels of government, and encourage industry to embrace a common vision, including through participation in existing voluntary programs.
  - EPA will take steps to bolster its voluntary Natural Gas STAR Program, which has already identified over 50 cost-effective technologies and practices that reduce or avoid methane emissions, by eliciting more robust industry commitments while enhancing transparency and accountability. In the spring of 2014, EPA will begin to engage the industry, states, and other key stakeholders on ways to enhance this program, and will formally launch the new partnership by the end of 2014.
- *Minimizing Venting and Flaring on Public Lands*: DOI's Office of Inspector General and the U.S. Government Accountability Office have both criticized BLM's outdated requirements governing venting and flaring for wasting Federal gas resources and associated royalties to the American taxpayer. To reduce the loss of natural gas through the venting or flaring of methane produced from Federal and Indian oil and gas leases, the BLM will develop a draft rule, known informally as Onshore Order 9, and anticipates releasing this proposed

rule later this year. To aid in the development of the rule, DOI has begun outreach to tribes, industry and other stakeholders.

Reducing Downstream Emissions

Safety is a top priority for natural gas distribution companies and state regulators. Thirty-eight states have some form of accelerated infrastructure replacement cost recovery program in place. Many companies are working with their state regulators to accelerate the modernization, replacement and expansion of the nation's natural gas pipeline system. These efforts to enhance safety also put people to work and reduce methane emissions. Several cooperative efforts are also underway. A group of 13 American Gas Association members are working with the Environmental Defense Fund on a project to improve measurement of methane emissions from natural gas distribution systems. In addition, the Natural Gas Downstream Initiative, a group of natural gas utilities, is collaborating to address key technical and regulatory factors affecting methane emission reduction opportunities from natural gas distribution systems. Through the initiative, partners are working to identify and encourage programs that accelerate investments in infrastructure and promote outstanding operations, including modernizing their systems and utilizing next generation technologies. The initiative is focused on opportunities that can substantially reduce methane emissions and support safe, reliable and cost-effective service. Current partners include Consolidated Edison Company of New York, National Grid, Pacific Gas & Electric Company, Public Service Electric and Gas Company, and Xcel Energy.

- Identifying Policy Recommendations for Reducing Emissions from Energy Infrastructure: The first installment of the Quadrennial Energy Review (QER), to be released in January of 2015, will recommend actions that industry, and Federal and state governments can take to improve the performance of our energy transmission, storage and distributions systems. Building on the DOE roundtables, the QER will evaluate methane emissions abatement opportunities from the processing, transmission, storage and distribution segments of the natural gas supply chain. To help identify the most cost-effective mitigation options, DOE's Office of Energy Policy and Systems Analysis will work with the National Labs and EPA to evaluate technology cost estimates developed by NGOs and industry, and combine analysis with associated emissions data, recently updated by EPA.

- Supporting Development of New Technologies to Reduce Emissions: DOE will support the development of new technologies to enable more cost-effective emission reductions through several programs:
    - In December 2013, the DOE made up to $8 billion in loan guarantee authority available for a wide array of advanced fossil energy projects under its Section 1703 loan guarantee program. Innovative technologies to reduce methane emissions from the coal mining and oil and gas sectors is one specific focus of this initiative, which will include regular solicitations for new loan applications.
    - The 2015 Budget proposes a new $4.7 million DOE program to speed development of technologies for leak detection and monitoring, pipeline leak repair without having to evacuate gas from the pipelines, smart pipeline sensors, and compressor controls. The program will be aimed at accelerating the commercialization of advanced pipeline inspection technologies. For example smart sensors could be distributed within the pipeline network and provide real-time continuous tracking of gas volumes and pipeline internal conditions.
- *Continuing to Prioritize Pipeline Safety:* The Pipeline and Hazardous Materials Safety Administration (PHMSA) will continue monitoring natural gas pipeline systems for safety, including requiring pipeline operators to take steps to eliminate leaks and prevent accidental releases of methane. Through this effort PHMSA has conducted a survey to evaluate each states progress to replace old, high-risk cast iron pipelines in their systems with new technologies that reduce leakage and accidental releases of methane over time.

U.S. Leadership in Reducing Global Methane Emissions

Methane accounts for approximately 15 percent of global annual greenhouse gas emissions. The United States is the recognized global leader in helping partners around the world to reduce methane emissions, including through the following two key actions:

- Spearheading Key Initiatives in the Climate and Clean Air Coalition: As a founding partner of the Climate and Clean Air Coalition to Reduce Short-Lived Climate Pollutants (CCAC) – which now includes nearly 40 country partners and key actors like the World Bank, UN Environment Programme, and World Health Assembly – the United States is helping to spearhead initiatives to across key sectors:

- The CCAC Municipal Solid Waste Initiative is currently working with 26 cities in Africa, Asia and Latin America to develop sustainable municipal solid waste practices to reduce methane emissions from landfills and improve air quality, public health, and the environment. To scale-up and replicate these efforts, the initiative is helping to build the capacity of national governments; has created a global city network that promotes peer-to-peer learning and sharing of best practices; and is developing innovative, sustainable financing solutions.
- The recently-launched CCAC Agriculture Initiative will promote improved manure management and rice cultivation practices through new global knowledge platforms, regional centers providing targeted assistance, and support to national and local early adopters of policies and technologies to catalyze large-scale practice change. This effort will not only reduce methane emissions and local pollution from the agriculture sector, it will also increase food security and productivity.
- The United States is working with international and corporate partners to launch the CCAC Oil and Gas Methane Partnership in 2014 with an initial group of oil and gas companies agreeing to systematically survey, report, and reduce methane emissions across a range of their participating operations. Participating companies will deploy proven and cost-effective technologies and practices across the largest sources of methane emissions in the oil and gas sector.

- Leveraging U.S. Technical Expertise through the Global Methane Initiative: Through the Global Methane Initiative (GMI), the United States leverages U.S. technical expertise to help partners around the world substantially reduce their methane emissions in five key sectors: agriculture; coal mines; municipal solid waste; oil and gas systems; and municipal wastewater. GMI is a public-private initiative with 43 partner countries, including all of the top 10 leading methane emitters, and over 1,000 members of the Project Network, including private sector, nongovernmental organizations, and multilateral organizations such as the World Bank, the Asian Development Bank, and the Inter-American Development Bank. GMI advances cost-effective, near-term methane recovery and use as a clean energy source, with a proven track record of helping to reduce methane emissions of over 220 million metric tons carbon dioxide since 2004, and identifying additional potential reductions that can be achieved cost-effectively. GMI has also developed a suite of tools and resources to help overcome barriers to methane capture and recovery and has built institutional capacity in Partner countries to ensure the long-term success of these efforts. These projects reduce greenhouse gas emissions in the near term and provide important environmental, safety, and economic co-benefits. The EPA is the lead agency from the U.S. Government and coordinates with the Department of State, Department of Energy, Department of Agriculture, USAID, and the Trade and Development Agency.

NGO-Industry Cooperative Research Initiative

In 2012, Environmental Defense Fund kicked off a series of studies — collaborating with more than 90 academic, research and industry partners — to better understand how much and from where methane is lost from the natural gas system today.

This project is investigating emissions from five key areas that make up the natural gas supply chain: production, gathering lines and processing facilities, long-distance pipelines and storage, local distribution and commercial trucks and refueling stations. The initiative includes 16 independent projects, all expected to be completed by the end of 2014.

III. Improving Measurement of Methane Sources and Emissions

Our current understanding of methane sources and trends supports the steps to reduce emissions outlined in this strategy. At the same time, sharpening our ability to measure emissions will enable more targeted efforts in the future. For this reason, in addition to identifying technologies and best practices for reducing emissions, the Climate Action Plan also calls for an assessment of current methane emissions data. Unlike carbon dioxide, where emissions are easily estimated from well-tracked energy statistics, many sources of methane are more diffuse, not systematically tracked by statistics agencies, and thus considerably more uncertain.

Methods for Measuring Methane Emissions

The measurement and estimation of methane emissions currently occurs in two primary ways: through "bottom-up" greenhouse gas inventories, which focus on the specific source or activity causing the emissions, and through "top-down" methods that infer emissions from measurements of atmospheric methane concentrations. Each approach has different strengths, weaknesses, and uncertainties, and they play complementary roles. Bottom-up inventories provide the foundation for policy and programs and top-down approaches are utilized for independent validation and overall assessments of the efficacy of national and international efforts to reduce methane in the atmosphere.

Nationally, emissions are tracked by the EPA through a bottom-up inventory, the U.S. Greenhouse Gas Inventory (GHGI), a document all parties to the UN Framework Convention on Climate Change (UNFCCC) are required to produce. A suite of methods for such inventories are published by the Intergovernmental Panel on Climate Change and generally consist of statistical approaches involving activity factors (*e.g.*, number of gas wells, number of landfills), emissions factors (*e.g.*, methane emissions per gas well, methane emissions per landfill), and reductions data (*e.g.*, counts of devices or practices that reduce methane emissions, such as flares and anaerobic digesters). The quality of methane data for some sources in the GHGI can be highly variable, and consequently, emissions estimates for some sources entail considerable uncertainty.

The GHGI is also beginning to be supplemented by new data from the Greenhouse Gas Reporting Program (GHGRP), a congressionally mandated EPA program requiring large emitters of GHGs from many sectors to estimate and report their emissions to EPA. Methane data are now available for several important sectors (*i.e.*, landfills, petroleum and natural gas systems, underground coal mines, and industrial wastewater systems) and are generally consistent with the GHGI, although more analysis is planned.

In addition to estimating human-related emissions using bottom-up statistical approaches, aggregate emissions can be inferred using top-down atmospheric models and measured concentrations of methane in the atmosphere. Atmospheric methane concentrations are sampled in the United States, as well as globally, by a variety of instruments on towers, ships, and aircraft. The network is adequate to estimate average global emissions, but it lacks the density to quantify emissions in all regions or to systematically identify emissions sources.

In sum, bottom-up methods are necessary to characterize emissions sources with precision, which is critical for designing mitigation strategies.

Top-down methods, while they cannot generally perform such attribution with high confidence, can help to validate bottom-up estimates using measured values and can help to identify emissions 'hot spots' for closer measurement.

Key Actions to Improve Methane Emissions Measurement and Monitoring

Administration efforts to improve U.S. methane measurement support two broad goals: 1) improving the bottom-up emissions data relevant for mitigation; and, 2) advancing the science and technology for monitoring and validating atmospheric concentrations. Within these broad goals, improvement opportunities exist across input data (*i.e.*, emissions factors, activity factors, and reductions data), atmospheric observations data, and the science needed to bridge between atmospheric observations and bottom-up emissions data (*i.e.*, monitoring and validation science). Federal agencies are already investing in related enhancements, and this strategy announces several critical new activities to further improve methane emissions measurement Examples of both include:

- *Encouraging the Development of Cost-Effective Measurement Technologies:* DOE's ARPA-E program is preparing a new methane program that will fund technologies to deliver an order-of-magnitude reduction on the cost of methane sensing, thus facilitating much wider deployment throughout all segments of natural gas systems.
- *Enhancing the US Greenhouse Gas Inventory*: EPA will continue to update and enhance the data published in its annual GHGI as new scientific evidence and data sources emerge. EPA will also continue to use the data collected through the GHGRP to improve the GHGI, particularly for the petroleum, natural gas, coal mining, and landfill sectors. This data will improve as additional reporting of inputs to emissions equations began in 2013 and, in 2015, EPA also plans to make ongoing improvements to the GHGRP regulatory requirements for petroleum and natural gas systems. In March 2014, EPA proposed revisions to GHGRP calculation methods, monitoring and data reporting requirements that would enhance the clarity and consistency of the reported data from petroleum and natural gas systems, such as for liquids unloading, completions and workovers, and compressors. The EPA will continue to review regulatory requirements to address potential gaps in coverage, improve methods, and help ensure high quality data reporting. DOE and USDA will also provide support to

improve emissions factors. EPA efforts to improve the GHGI will promote transparency and stakeholder input by means of annual expert, public and international review periods.

- *Building our National Methane Monitoring Network:* National Oceanic and Atmospheric Administration (NOAA) scientists maintain a network of methane monitoring sites in the United States, including tall towers, periodic aircraft measurements, and surface measurements. NOAA has also conducted periodic aircraft-based methane measurements in six major U.S. oil and gas production regions. At its current funding level, this Carbon Observation and Analysis Program provides the minimum needed for climate modeling. To expand capabilities, the President's budget requests $8 million above current funding of $6.5 million for this program to:
  - Add 6 tall towers to the network, increasing the network to 14;
  - Enhance the measurement capabilities of all 14 towers;
  - Triple the frequency of aircraft-based observations.
- Improving Local & Regional Emissions Modeling: As part of DOE's ongoing unconventional gas program, DOE is funding two projects - one at Pennsylvania State University and one at Carnegie Mellon University - using tracer release methods and tower, automobile, aircraft monitoring, and other methods to measure and model methane emissions from the Marcellus region in Pennsylvania. A regional inventory of other methane sources including landfills, wetlands, water treatment facilities, and agriculture sources will also be obtained. The project is scheduled to begin in 2015 and end in 2017. Additionally, NASA's Jet Propulsion Laboratory JPL, is carrying out a Carbon in Arctic Reservoirs Vulnerability Experiment, which includes regular monitoring of methane concentrations over Alaska's North Slope.
- Improving Global Emissions Monitoring and Estimates: EPA is collecting emissions reduction data through the Global Methane Initiative. EPA will also continue to update and publish detailed estimates and projections of global human-related non-CO2 greenhouse gas emissions, and the mitigation potential from these sources. DOE's Atmospheric Radiation Measurement Climate Research Facility is making long-term methane flux measurements at multiple permanent locations around the world. NOAA runs the largest global network of GHG measurements and works closely with international partners and the World Meteorological Organization to

ensure global measurements of GHG concentrations, including methane, are standardized. NOAA also consolidates data from this global network and releases the data to the public. Other Federal agencies (e.g. NASA and DOE) also contribute to these networks. NASA and the National Institute of Standards and Technology are helping to fund the Megacities Carbon Project, an international research effort to develop and demonstrate a scientifically robust capability to measure multi-year emission trends of CO2, methane, and carbon monoxide attributed to individual megacities and selected major sectors in those cities. In addition, USDA's Forest Service is working with international partners, universities and the USAID on international efforts that monitor methane on sites in Mexico, Ecuador, Colombia, and Peru.

End Notes

[1] Estimates of methane's potency as a greenhouse gas in this document use 100-year global warming potential values from the Second Assessment Report of the Intergovernmental Panel on Climate Change, as required by international reporting standards.

[2] These estimates are based on the Draft Inventory of U.S. Greenhouse Gas Emissions and Sinks: 1990-2012, EPA, February 2014. The draft 2014 GHG Inventory (calculating emissions from 1990 through 2012) includes several updates to the methane numbers. As a result of the recalculations, in the draft 2014 Inventory, the total methane emissions estimate decreased by about 4 percent from the previous estimate.

In: Methane
Editor: Andrea Malone

ISBN: 978-1-63482-526-9

Chapter 4

U.S. METHANE EMISSIONS 1990-2020: INVENTORIES, PROJECTIONS, AND OPPORTUNITIES FOR REDUCTIONS*

U.S. Environmental Protection Agency

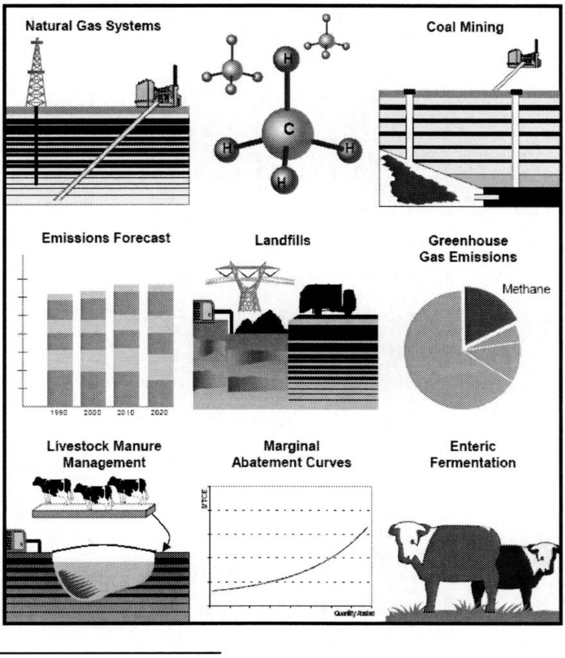

* This is an edited, reformatted and augmented version of a document, EPA 430-R-99-013, issued by the U.S. Environmental Protection Agency, September 1999.

ABBREVIATIONS, ACRONYMS, AND UNITS

AF	Activity factor
ASAE	American Society of Agricultural Engineers
Bcf	Billion cubic feet
BMP	Best management practice
CAA	Clean Air Act
CCAP	Climate Change Action Plan
C&D	Construction and demolition
CFC	Chlorofluorocarbon
CH4	Methane
CMOP	Coalbed Methane Outreach Program
CO2	Carbon dioxide
DI&M	Directed inspection and maintenance
DOE	Department of Energy
EF	Emission factor
EIA	Energy Information Administration
EPA	Environmental Protection Agency
E-PLUS	Energy Project Landfill Gas Utilization Software
GAA	Government Advisory Associates
GHG	Greenhouse gas
GSAM	Gas Systems Analysis Model
GWP	Global warming potential
IC	Internal combustion
IPCC	Intergovernmental Panel on Climate
kW	kilowatt
kWh	kilowatt-hour
LMOP	Landfill Methane Outreach Program
MAC	Marginal abatement curve
Mcf	Thousand cubic feet
MMBtu	Million British thermal units
MMcf/d	Million cubic feet per day
MMTCE	Million (metric) tons of carbon equivalent
MMT	Million (metric) tons
MSHA	Mine Safety and Health Administration
MSW	Municipal solid waste
MW	Megawatt

NMOC	Non-methane organic compound
NPV	Net present value
O&M	Operation and maintenance
PRO	Partner-reported opportunity
RLEP	Ruminant Livestock Efficiency Program
Tcf	Trillion cubic feet
Tg	Teragram
TCE	Metric ton of carbon equivalent
UNFCCC	United Nations Framework Convention on Climate Change
USDA	United States Department of Agriculture
VOC	Volatile organic compound
WIP	Waste-in-place

CONVERSIONS

1 Mcf Methane = 1 MMBtu
1 Bcf = 1,000 MMcf
1 Tg = 1×10^{12} g
1 Tg CH_4 = 1 MMT CH_4
1 MMT CH_4 = 5.73 MMTCE
GWP of CO_2 = 1
GWP of CH_4 = 21

EXECUTIVE SUMMARY

Methane gas is a valuable energy resource and the leading anthropogenic contributor to global warming after carbon dioxide. Atmospheric methane concentrations have doubled over the last 200 years and continue to rise, although the rate of increase is slowing (Dlugokencky, et al., 1998). By mass, methane has 21 times the global warming potential of carbon dioxide over a 100-year time frame. Methane accounts for 10 percent of U.S. greenhouse gas emissions (excluding sinks) and reducing these emissions is a key goal of the U.S. Climate Change Action Plan (EPA, 1999).

The major sources of anthropogenic methane emissions in the U.S. are landfills, agriculture (livestock enteric fermentation and manure management), natural gas and oil systems, and coal mines. Smaller sources in the U.S. include rice cultivation, wastewater treatment, and others. Unlike other greenhouse gases, methane can be used to produce energy since it is the major component (95 percent) of natural gas. Consequently, for many methane sources, opportunities exist to reduce emissions cost-effectively or at low cost by capturing the methane and using it as fuel.

This report has two objectives. First, it presents the U.S. Environmental Protection Agency's (EPA's) baseline forecast of methane emissions from the major anthropogenic sources in the U.S., and EPA's cost estimates of reducing these emissions. Emission estimates are given for 1990 through 1997 with projections for 2000 to 2020. The cost analysis is for 2000, 2010, and 2020. Second, this report provides a transparent methodology for the calculation of emission estimates and reduction costs, thereby enabling analysts to replicate these results or use the approaches described herein to conduct similar analyses for other countries.

Baseline Methane Emission Estimates

EPA estimates annual emissions for 1990 to 1997 and forecasts emissions for 2000, 2010, and 2020. In 1990, the U.S. emitted 169.9 million metric tons of carbon equivalent (MMTCE) or 29.7 Teragrams (Tg) of methane. By 1997, estimated methane emissions were slightly higher at 179.6 MMTCE (31.4 Tg) (EPA, 1999). The baseline U.S. methane emission forecast for 2010 is 186.0 MMTCE (32.5 Tg) which is almost a ten percent increase over the 1990 levels. However, this forecast excludes the expected reductions associated with U.S. voluntary programs. When these programs are taken into account, methane emissions are expected to remain at or below 1990 levels through 2020. Exhibit ES-1 shows current methane emissions and projections by industry.

To estimate historic and future emissions, EPA characterizes the source industries in detail and identifies the specific processes within those industries that produce emissions. Forecasts are based on a consistent set of industry factors, e.g., consumption, prices, technological change, and infrastructure makeup. The major emission sources are outlined below.

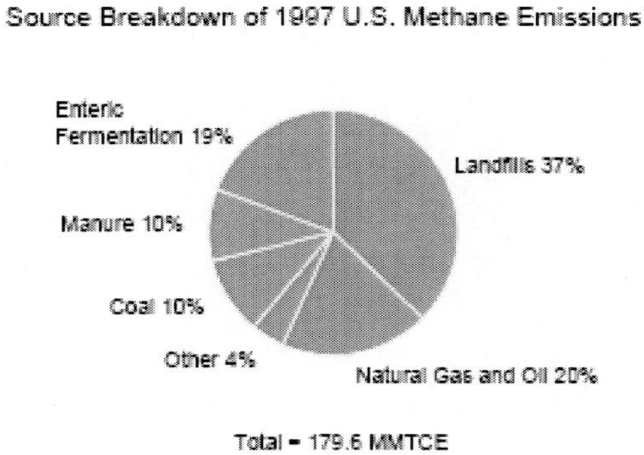

Source: EPA,1999.

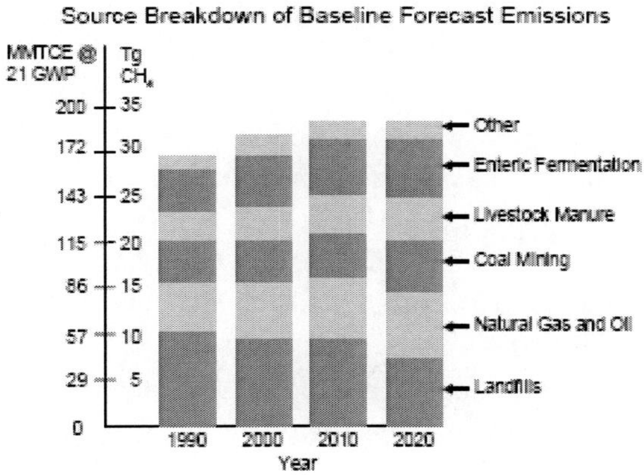

Exhibit ES-1. U.S. Methane Emissions (MMTCE).

– *Landfills.* The largest source (accounting for 37 percent) of U.S. anthropogenic methane emissions, landfills generate methane during anaerobic decomposition of organic waste. In 1990, landfills generated 56.2 MMTCE (9.8 Tg) of methane, which increased to 66.7 MMTCE (11.6 Tg) by 1997 (EPA, 1999). Baseline emissions are expected to decrease to 52.0 MMTCE (9.1 Tg) in 2010, due to the Clean Air Act New Source Performance Standards and Emissions Guidelines (Landfill Rule). The Landfill Rule requires the nation's largest landfills to reduce emissions of non-methane organic compounds and results in a simultaneous reduction in methane emissions. The principal technologies for reducing emissions from landfills involve collecting methane and using it as fuel for electric power generation or for sale to nearby industrial users.

- *Natural Gas Systems.* Emissions of methane occur throughout the natural gas system from leaks and venting of gas during normal operations, maintenance, and system upsets. In 1990, methane emissions from the U.S. natural gas system totaled about 32.9 MMTCE (5.7 Tg), and by 1997 methane emissions were estimated at 33.5 MMTCE (5.8 Tg) (EPA, 1999). EPA expects emissions to increase as natural gas consumption increases, although at a lower rate than gas consumption growth. Baseline emissions reach 37.9 MMTCE (6.6 Tg) in 2010. Improved management practices and technologies can reduce leaks or avoid venting of methane from all parts of the natural gas system.
- *Coal Mining.* Methane and coal are formed together by geological forces during coalification. As coal is mined, the methane is released. Because methane is hazardous to miners, underground mines use ventilation systems to dilute it and additional techniques to recover it during or in advance of mining. In 1990, coal mine methane emissions were estimated at 24.0 MMTCE (4.2 Tg). By 1997, emissions fell to 18.8 MMTCE (3.3 Tg) mainly due to reduced coal production at "gassy" mines and increased methane recovery (EPA, 1999). Baseline methane emissions reach 28.0 MMTCE (4.9 Tg) by 2010 due to growth in coal mining from deep mines. The major technologies for reducing emissions include recovery and sale to pipelines, use for power generation, or on-site use. Catalytic oxidation of methane in ventilation air may also be undertaken to reduce emissions.
- *Livestock Manure Management.* Methane is produced during the anaerobic decomposition of livestock manure. The major sources of U.S. livestock manure methane include large dairy and cattle operations and hog farms that use liquid manure management systems. In 1990, livestock manure emitted about 14.9 MMTCE (2.6 Tg) of methane. Emissions from this source increased to 17.0 MMTCE (3.0 Tg) by 1997 (EPA, 1999). Baseline emissions reach 22.3 MMTCE (3.9 Tg) in 2010 due to animal population growth driven by increases in total meat and dairy product consumption and increasing use of liquid waste management systems that produce methane. Existing cost-effective technologies can be used to recover this methane to produce energy.
- *Enteric Fermentation.* Methane emissions from livestock enteric fermentation were 32.7 MMTCE (5.7 Tg) in 1990 and 34.1 MMTCE (6.0 Tg) in 1997 (EPA, 1999). Baseline methane emissions reach 37.7

MMTCE (6.6 Tg) by 2020 due to increased domestic and international demand for U.S. livestock products. Emissions can be reduced through the application of improved management practices. The cost-effectiveness of these practices has not been quantified as part of this analysis, however.

COSTS OF REDUCING EMISSIONS

This report presents the results of extensive benefit-cost analyses conducted on the opportunities (technologies and management practices) to reduce methane emissions from four of the five major U.S. sources: landfills, natural gas systems, coal mining, and livestock manure. To date, most economic analyses of U.S. greenhouse gas (GHG) emission reductions have focused on energy-related carbon emissions since carbon dioxide (CO_2) currently accounts for about 82 percent of the total U.S. GHG emissions (weighted by 100-year global warming potentials) (EPA, 1999). The cost estimates for reducing methane emissions presented in this report can be integrated into economic analyses to produce more comprehensive assessments of total GHG reductions. By including methane emission reductions, the overall cost of reducing GHG emissions in the U.S. is reduced. At increasing values for emission reductions, in terms of dollars per metric ton of carbon equivalent ($/TCE), more costly CO_2 reductions can be substituted by lower cost methane reductions, when available, thereby lowering the marginal cost and the total cost of a particular GHG emission reduction level.

The cost analysis is conducted for the years 2000, 2010, and 2020. All values are in 1996 constant dollars. Results for the source-specific analyses are summarized below.

- *Landfills.* The cost analysis focuses on technologies for recovering and using landfill methane for energy. Two options are evaluated: use of landfill methane for electricity generation and as a fuel for direct use by a nearby end-user. After accounting for emission reductions due to the Landfill Rule, at $0/TCE, about 21 percent of baseline emissions from landfills could be captured and used cost-effectively in 2000. Cost-effective reductions decrease slightly to 20 percent, at $0/TCE, in 2010, in part reflecting greater coverage of total emissions by the Landfill Rule. At $30/TCE, emissions could be reduced by 38 percent from the baseline in 2000, and by 41 percent in 2010.

Emission reductions approach their maximum at $100/TCE in 2000, and $40/TCE in 2010. EPA projects the incremental benefits of higher values for carbon equivalent to be slightly smaller in 2020 due to the Landfill Rule.

- *Natural Gas Systems.* Cost curves for reducing methane emissions from natural gas systems are based on technologies and practices for reducing leaks and venting of natural gas in the natural gas system. EPA evaluates 118 technologies and practices that have been identified by the gas industry in conjunction with EPA's Natural Gas STAR Program. EPA's analysis assesses the cost-effectiveness of each technology and practice based on the value of methane as natural gas. In 2000, 2010, and 2020, about 30 percent of the projected emissions from natural gas systems can be avoided cost-effectively, based on the value of the saved methane. When a value of $30/TCE for avoided emissions is added to the market price for gas, about 35 percent of the emissions can be reduced. At $100/TCE, about 49 percent of emissions can be reduced. Additional technologies could likely emerge in this sector to reduce emissions at high values for carbon equivalent, however, EPA only examines current technologies in this analysis.
- *Coal Mining.* EPA's analysis for reducing coal mine methane emissions focuses on recovering methane from underground mining, which comprises 65 percent of the emissions from this source. Two emission reduction strategies are analyzed: recovering methane from mines for sale as natural gas and using new catalytic oxidation technologies. The results suggest that in 2010, 37 percent of emissions from coal mines can be cost-effectively reduced at energy market prices, or $0/TCE. Up to 71 percent of emissions can be reduced at $30/TCE, which represents essentially all of the technically recoverable methane from this source. In 2020, the same pattern exists with 41 percent recoverable at $0/TCE and 71 percent recoverable at $30/TCE.
- *Livestock Manure Management.* Cost curves for reducing methane emissions from livestock manure are based on recovering and utilizing methane produced at dairies and swine farms. EPA's analysis focuses on anaerobic digestion technologies (including covered and complete mix digesters) that capture methane for use on-site to generate electricity. At current energy prices, emissions from livestock manure could be reduced by 14 percent in 2000 and 2010.

Emission reductions increase slightly to 15 percent in 2020. With an additional \$30/TCE, emission reductions reach 30 percent in 2000, 31 percent in 2010, and 32 percent in 2020. At \$100/TCE, emissions can be reduced by about 63 percent in 2000, 65 percent in 2010, and 67 percent in 2020.

– *Enteric Fermentation.* Emissions from livestock enteric fermentation can be reduced through enhanced feeding and animal management techniques. The costs and cost-effectiveness of these reductions have not been quantified for this report.

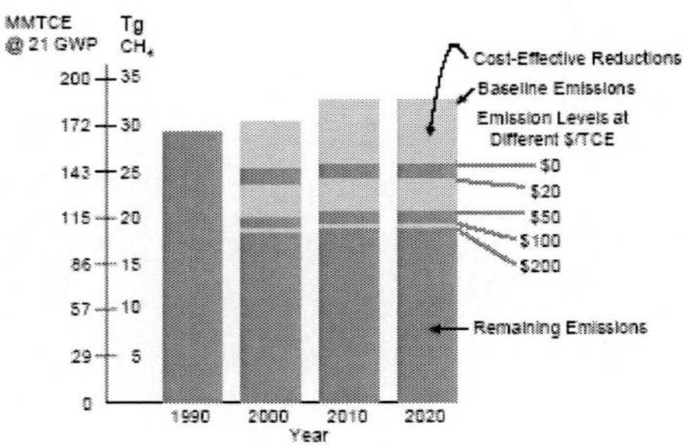

	2000	2010	2020
Baseline Emissions	173.9	186.0	183.7
Cumulative Reductions			
at \$0/TCE	30.8	34.8	35.0
at \$10/TCE	36.4	42.3	40.9
at \$20/TCE	41.7	50.3	47.4
at \$30/TCE	54.6	61.7	58.7
at \$40/TCE	56.2	63.5	61.0
at \$50/TCE	59.5	66.9	64.8
at \$75/TCE	64.3	71.9	70.7
at \$100/TCE	67.2	74.9	74.0
at \$125/TCE	68.4	76.2	75.5
at \$150/TCE	68.7	76.5	75.9
at \$175/TCE	69.0	76.8	76.2
at \$200/TCE	69.2	77.0	76.5
Remaining Emissions	104.7	108.9	107.2

Exhibit ES-2. U.S. Baseline Emissions and Potential Reductions (source-specific discount rates) (MMTCE).

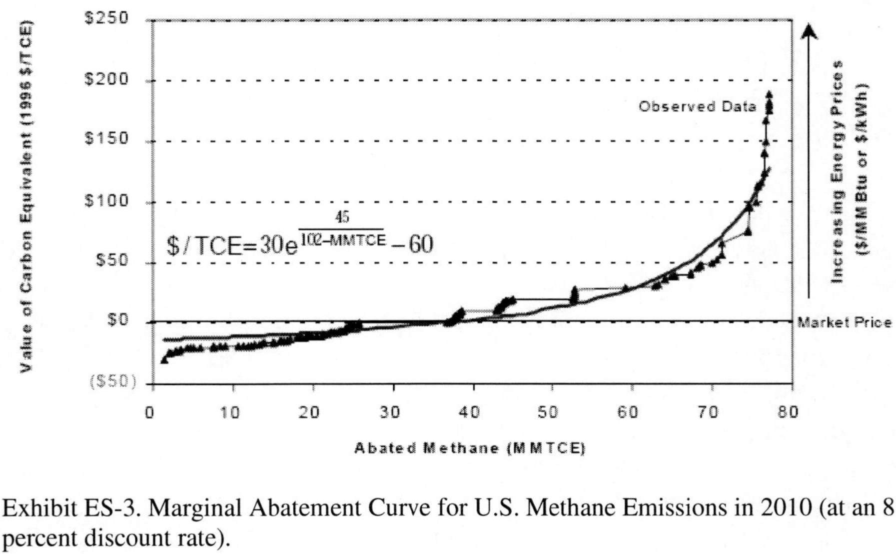

Exhibit ES-3. Marginal Abatement Curve for U.S. Methane Emissions in 2010 (at an 8 percent discount rate).

$20/TCE and $50/TCE total estimated reductions are 52.6 MMTCE (9.2 Tg) and 70.0 MMTCE (12.2 Tg), respectively. Third, at $100/TCE, total achievable reductions are estimated at 75.5 MMTCE (13.2 Tg). Finally, above $100/TCE, the MAC becomes inelastic, that is, non-responsive to increasing methane values. This inelasticity indicates the limits of the options considered. The magnitude of the cost-effective and low-cost reductions reflects methane's value as an energy source and emphasizes that many proven technologies can be used to recover it. For several sources, the inelastic section of the curve at the higher end of the cost range indicates a limitation of the analysis, namely that only available technologies are assessed. Additional technologies may become available to reduce methane emissions at these prices; however, EPA has not yet assessed this possibility.

EPA has developed a number of voluntary programs as part of the Climate Change Action Plan (CCAP) to overcome market barriers and encourage cost-effective methane recovery projects. In this report, the emission reductions associated with these CCAP programs have not been subtracted from the baseline emission projections.

However, EPA expects that approximately 50 percent of the reductions available in 2010 at $0/TCE will be captured by these programs. These programs have reduced emissions by 8 MMTCE in 1998 and are expected to reduce emissions by 12 MMTCE in 2000, and 20 MMTCE in 2010.

1. Introduction and Aggregate Results

Introduction

This report has two objectives. First, it presents the U.S. Environmental Protection Agency's (EPA's) baseline forecast of methane emissions from the major anthropogenic sources in the U.S., and EPA's cost estimates of reducing these emissions. Emission estimates are given for 1990 through 1997 with projections for 2000 to 2020. The cost analysis is for 2000, 2010, and 2020. Second, this report provides a transparent methodology for the calculation of emission estimates and reduction costs, thereby enabling analysts to replicate these results or use the approaches described herein to conduct similar analyses for other countries.

The information presented in this report can be used in several ways. The emission estimates and forecasts represent the most up-to-date estimates of methane emissions in the U.S.; thus, this report replaces and expands upon

EPA's *Anthropogenic Methane Emissions in the United States, Estimates for 1990, Report to Congress* (1993a). As such, this report can be used where estimates of future emissions are required. The report also summarizes the state of knowledge on methane emissions from the major anthropogenic sources.

While the emission estimations are refinements of earlier approaches, the cost analyses presented in this report represent a major contribution to the literature on mitigating emissions. To date, most economic analyses of greenhouse gas (GHG) emission reductions have focused on the energy-related carbon emissions since carbon dioxide (CO_2) currently accounts for about 82 percent of the total U.S. emissions (weighted by 100-year global warming potentials) (EPA, 1999). The cost-estimates for reducing methane emissions presented in this report can be integrated into economic analyses to produce more comprehensive assessments of total GHG reductions. By including methane emission reductions, the overall cost of reducing GHG emissions in the U.S. is reduced. At increasing values for emission reductions, more costly CO_2 reductions can be substituted by lower cost methane reductions, when available, thereby lowering the marginal cost and the total cost of a particular GHG emission reduction level.

The marginal abatement curves (MACs) developed in this report can be used to estimate possible emission reductions at various prices for carbon equivalent emissions or conversely, the costs of achieving certain amounts of reductions. EPA recognizes that the cost analyses will change with the introduction of new technologies and additional research into methane emission abatement technologies. Other countries, nevertheless, can use the cost analyses presented in this report as the basis for estimating emission reduction costs.

1.0. Overview of Methane Emissions

Next to carbon dioxide, methane is the second largest contributor to global warming among anthropogenic greenhouse gases. Methane's overall contribution to global warming is significant because, over a 100-year time frame, it is estimated to be 21 times more effective at trapping heat in the atmosphere than carbon dioxide. As illustrated in Exhibit 1-1, methane accounts for 17 percent of the enhanced greenhouse effect (IPCC, 1996a).[1]

Over the last two centuries, methane's concentration in the atmosphere has more than doubled from about 700 parts per billion by volume (ppbv) in preindustrial times to 1,730 ppbv in 1997 (IPCC, 1996a). Exhibit 1-1 illustrates this trend. Scientists believe these atmospheric increases are largely due to increasing emissions from anthropogenic sources. Although atmospheric methane concentrations continue to rise, the rate of increase appears to have slowed since the 1980s. If present trends continue, however, atmospheric methane concentrations will reach 1,800 ppbv by 2020 (Dlugokencky, et al., 1998).

Atmospheric methane is reduced naturally by sinks. Natural sinks are removal mechanisms and the greatest sink for atmospheric methane (CH_4) is through a reaction with naturally-occurring tropospheric hydroxyl (OH).[2] Methane combines with OH to form water vapor (H_2O) and carbon monoxide (CO), which in turn is converted into carbon dioxide (CO_2). Atmospheric methane, nevertheless, has a clearly defined chemical feedback that decreases the effectiveness of the hydroxyl sink. As methane concentrations rise, less hydroxyl is available to break down methane, producing longer atmospheric methane lifetimes and higher methane concentrations (IPCC, 1996a).

On average, the atmospheric lifetime for a methane molecule is 12.2 years (± 3 years) before a natural sink consumes it (IPCC, 1996a). This relatively short lifetime makes methane an excellent candidate for mitigating the impacts of global warming because emission reductions could lead to stabilization or reduction in methane concentrations within 10 to 20 years.

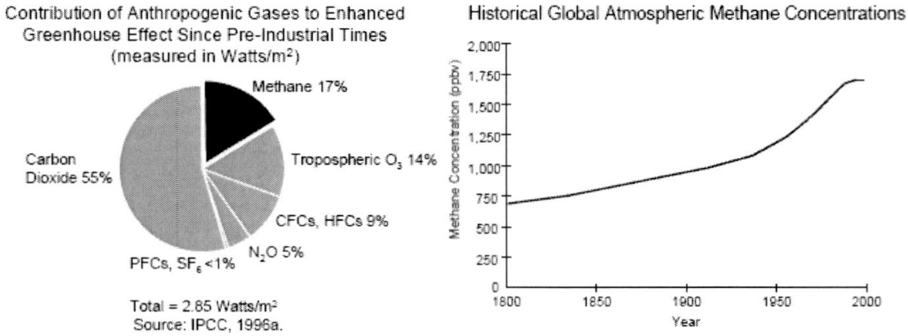

Source: Boden, et al., 1994; Dlugokencky, et al., 1998.

Exhibit 1-1. Global Enhanced Greenhouse Effect and Methane Concentrations.

2.0. Sources of Methane Emissions

Methane is emitted into the atmosphere from both natural and anthropogenic sources. Natural sources include wetlands, tundra, bogs, swamps, termites, wildfires, methane hydrates, and oceans and freshwaters. Anthropogenic sources include landfills, natural gas and oil production and processing, coal mining, agriculture (livestock enteric fermentation and livestock manure management, and rice cultivation), and various other sources. By 1990, anthropogenic sources accounted for 70 percent of total global methane emissions (EPA, 1993a; IPCC, 1996a). This section summarizes the natural and anthropogenic sources of methane.

2.1. Natural Methane Emissions

In 1990, worldwide natural sources emitted 916 million metric tons of carbon equivalent (MMTCE) or 160 Teragrams (Tg) of methane into the atmosphere, or about 30 percent of the total methane emissions (IPCC, 1996a). The leading natural methane sources are described below in descending order of their contribution to emissions (see Exhibit 1-2).

Wetlands. Methane is generated by anaerobic (oxygen poor) bacterial decomposition of plant material in wetlands. Natural wetlands emit about 659 MMTCE (115 Tg) of methane per year, which is 72 percent of natural emissions and 20 percent of total global methane emissions (IPCC, 1995). Methane emissions from wetlands will probably increase with global warming as a result of accelerated anaerobic microbial activity. In addition, climate change models predict increased precipitation as global temperatures rise, which could create more wetlands (EPA, 1993b). Tropical wetlands (between 20° N and 30° S) represent 17 percent of total wetland area and 60 percent of emissions from wetlands. These relatively high emissions are due to higher temperatures, more precipitation and more intense solar radiation, which encourage higher plant growth and decomposition rates (EPA, 1993b).

Northern Wetlands (those above 45° N) are usually underlain with near-surface permafrost that prevents soil drainage and creates wetland conditions. Northern wetlands represent nearly 80 percent of the wetland area and 35 percent of methane emissions from wetlands (EPA, 1993b).

Termites. Microbes within the digestive systems of termites break down cellulose, and this process produces methane. Emissions from this source depend on termite population, amounts of organic material consumed, species, and the activity of methane-oxidizing bacteria. While more research is needed, some experts believe that future trends in termite emissions are more influenced by anthropogenic changes in land use, i.e., deforestation for agriculture, than by climate change. Termites emit an estimated 115 MMTCE (20 Tg) of methane each year (IPCC, 1995).

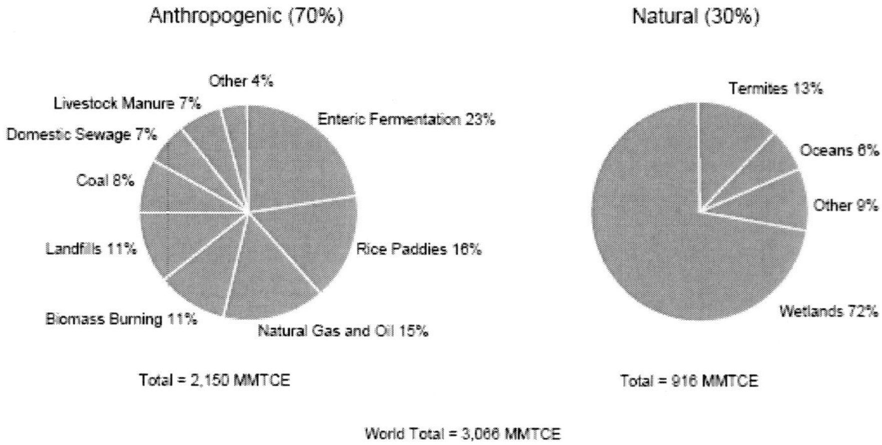

Source: IPCC, 1995 and 1996a.

Exhibit 1-2. Worldwide Natural and Anthropogenic Methane Emissions in 1990.

Oceans and Freshwaters. The surface waters of the world's oceans and freshwaters are slightly supersaturated with methane relative to the atmosphere and therefore emit an estimated 57 MMTCE (10 Tg) of methane each year (IPCC, 1995). The origin of the dissolved methane is not known. In coastal regions it may come from sediments and drainage. It also has been suggested that methane is generated in the anaerobic gastrointestinal tracts of marine zooplankton and fish (EPA, 1993b). Methane in freshwaters can result from the decomposition of wetland plants. (In this report, methane emissions from freshwaters are included in the estimates for wetlands.) As atmospheric methane concentrations increase, the proportion of methane supersaturated in oceans and freshwaters will decline relative to the atmospheric concentrations of methane, assuming that the methane concentration in oceans and freshwaters remains constant.

Gas Hydrates. Methane is trapped in gas hydrates, which are dense combinations of methane and ice located deep underground and beneath the ocean floor. Recent estimates of hydrates suggest that around 44 billion MMTCE (7.7 billion Tg) of methane is trapped in both oceanic and continental gas hydrates (DOE, 1998). Scientists agree that increasing temperatures will eventually destabilize many gas hydrates, but are unsure about the timing and the amount of methane emissions that would be released from the deeply buried hydrates (EPA, 1993b).

Permafrost. Small amounts of methane are trapped in permafrost, which consists of permanently frozen soil and ice. (To be classified as permafrost, the ice and soil mixture must remain at or below 0° Celsius year-round for at least two consecutive years.) Due to the large amount of existing permafrost, the total amount of methane stored in this form could be quite high, possibly several thousand Tg (EPA, 1993b). This methane is released when permafrost melts. However, no estimates have been made for current emissions from this source.

Wildfires. Wildfires are primarily caused by lightning and release a number of greenhouse gases, including methane which is a product of incomplete combustion. However, no estimates are available for methane emissions from this source.

2.2. Anthropogenic Methane Emissions

Methane emissions from anthropogenic sources account for 70 percent of all methane emissions and totaled 2,150 MMTCE (375 Tg) worldwide in 1990 (IPCC, 1996a). The leading global anthropogenic methane sources are described below in descending order of magnitude. The two leading sources of anthropogenic methane emissions worldwide are livestock enteric fermentation and rice production. By contrast, in the U.S., the two leading sources of methane emissions are landfills and natural gas and oil systems (see Exhibit 1-3). In 1997, the U.S. emitted 179.6 MMTCE (31.4 Tg) of methane, about 10 percent of global methane emissions for that year (EPA, 1999). The U.S. is the fourth-largest methane emitter after China, Russia, and India (EPA, 1994).

Enteric Fermentation. Ruminant livestock emit methane as part of their normal digestive process, during which microbes break down plant material consumed by the animal into material the animal can use. Methane is produced as a by-product of this digestive process, and is expelled by the animal. In the U.S., cattle emit about 96 percent of the methane from livestock enteric fermentation. In 1994, livestock enteric fermentation produced 490 MMTCE (85 Tg) of methane worldwide (IPCC, 1995), with the emissions coming from the former Soviet Union, Brazil, and India (EPA, 1994).

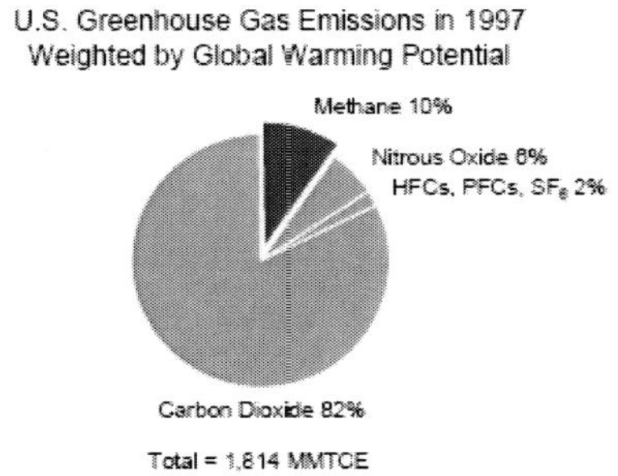

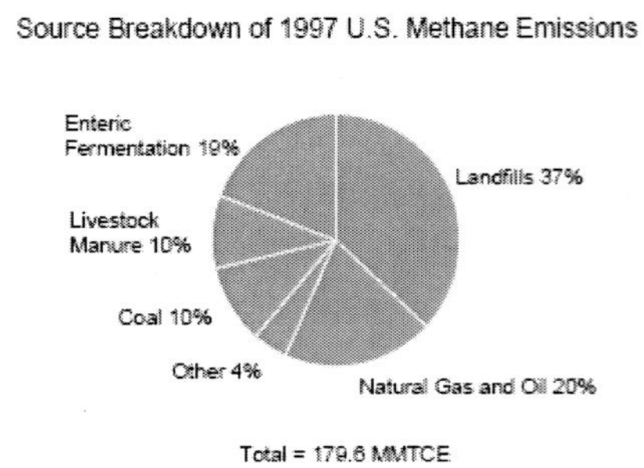

Source: EPA, 1999.

Exhibit 1-3. U.S. Methane Emissions.

Exhibit 1-4. Baseline Methane Emissions in the United States (MMTCE)

Source	1990[a]	1997[a]	2000	2010	2020
Landfills	56.2	66.7	51.4	52.0	41.1
Natural Gas Systems	32.9	33.5	35.6	37.9	38.8
Oil Systems	1.6	1.6	1.6	1.6	1.7
Coal Mining	24.0	18.8	23.9	28.0	30.4
Livestock Manure Management	14.9	17.0	18.4	22.3	26.4
Enteric Fermentation	32.7	34.1	35.2	36.6	37.7
Other[b]	7.3	7.4	7.8	7.6	7.6
Total	169.9	179.6	173.9	186.0	183.7

[a] Source: EPA, 1999.

[b] These estimates developed by EPA for the 1997 Climate Action Report (DOS, 1997). Totals may not sum due to independent rounding.

EPA estimates that U.S. emissions from this source were 34.1 MMTCE (6.0 Tg) in 1997 (EPA, 1999). Under EPA's baseline forecast, livestock enteric fermentation emissions in the U.S. will increase to about 37.7 MMTCE (6.6 Tg) by 2020 (Exhibit 1-4). The projected increase is due to greater consumption of meat and dairy products.

Rice Paddies. Most of the world's rice, including rice in the United States, is grown on flooded fields where organic matter in the soil decomposes under anaerobic conditions and produces methane. The U.S. is not a major producer of rice and therefore emits little methane from this source. Worldwide emissions of methane from rice paddies were 345 MMTCE (60 Tg) in 1994 (IPCC, 1995), with the highest emissions coming from China, India, and Indonesia (EPA, 1994). EPA estimates U.S. emissions from this source at 2.7

MMTCE (0.5 Tg) in 1997 and expects emissions to remain stable in the future (EPA, 1999).

Natural Gas and Oil Systems. Methane is the major component (95 percent) of natural gas. During production, processing, transmission, and distribution of natural gas, methane is emitted from system leaks, deliberate venting, and system upsets (accidents). Since natural gas is often found in conjunction with petroleum, crude petroleum gathering and storage systems are also a source of methane emissions. In 1994, natural gas systems worldwide emitted 230 MMTCE (40 Tg) of methane and oil systems emitted 85 MMTCE (15 Tg) of methane (IPCC, 1995). EPA estimates that 1997 U.S. emissions were 33.5 MMTCE (5.8 Tg) from natural gas systems and 1.6 MMTCE (0.27 Tg) from oil systems (EPA, 1999). EPA expects emissions from oil systems to remain near 1997 levels through 2020. The baseline emission forecast is 38.8 MMTCE (6.8 Tg) from natural gas systems in 2020 (Exhibit 1-4). The increase results from higher consumption of natural gas and expansions of the natural gas system.

Biomass Burning. Biomass burning releases greenhouse gases, including methane, but is not a major source of U.S. methane emissions. In 1994, biomass burning produced 230 MMTCE (40 Tg) of methane worldwide (IPCC, 1995). EPA estimates that U.S. emissions from this source were 0.2 MMTCE (0.03 Tg) in 1997 and that emissions will remain stable through 2020 (EPA, 1999).

Landfills. Landfill methane is produced when organic materials are decomposed by bacteria under anaerobic conditions. In 1994, landfills produced 230 MMTCE (40 Tg) of methane worldwide (IPCC, 1995). EPA estimates that U.S. emissions from this source were 66.7 MMTCE (11.6 Tg) in 1997 (EPA, 1999). The baseline forecast is 41.1 MMTCE (7.2 Tg) from U.S. landfills in 2020 (Exhibit 1-4). Landfill methane is the only U.S. source that is expected to decline in the baseline over the forecast period. This decline is due to the implementation of the New Source Performance Standards and Emissions Guidelines (the Landfill Rule) under the Clean Air Act (March 1996). While the Landfill Rule controls greenhouse gas emissions that form tropospheric ozone (smog), it also will lead to lower methane emissions. The Landfill Rule requires large landfills to collect and combust or use landfill gas emissions.

Coal Mining. Methane is trapped within coal seams and the surrounding rock strata and is released during coal mining. Because methane is explosive in low concentrations, underground mines install ventilation systems to vent methane directly to the atmosphere. In 1994, coal mining produced 170

MMTCE (30 Tg) of methane worldwide (IPCC, 1995). EPA estimates that U.S. emissions from this source were 18.8 MMTCE (3.3 Tg) in 1997 (EPA, 1999). EPA's baseline estimate indicates that emissions from coal mines could reach 30.4 MMTCE (5.3 Tg) by 2020 (Exhibit 1-4). The increase results from greater coal production from deep mines.

Domestic Sewage. The decomposition of domestic sewage in anaerobic conditions produces methane. Domestic sewage is not a major source of methane emissions in the U.S., where it is collected and processed mainly in aerobic (oxygen rich) treatment plants. In 1994, domestic sewage produced 145 MMTCE (25 Tg) of methane worldwide (IPCC, 1995). EPA estimates that emissions from sewage in the U.S. were 0.9 MMTCE (0.2 Tg) in 1997 and expects emissions to increase only slightly by 2020 (EPA, 1999). This increase will be due primarily to population increases.

Livestock Manure Management. The decomposition of animal waste in anaerobic conditions produces methane. Over the last eight years, methane emissions from manure have generally followed an upward trend. This trend is driven by: (1) increased swine and poultry production; and (2) increased use of liquid manure management systems, which create the anaerobic conditions conducive to methane production. In 1994, manure management produced 145 MMTCE (25 Tg) of methane worldwide (IPCC, 1995). EPA estimates that U.S. emissions from this source were 17.0 MMTCE (3.0 Tg) in 1997 (EPA, 1999). Emissions from livestock manure in the baseline are projected to increase to 26.4 MMTCE (4.6 Tg) by 2020 (Exhibit 1-4) mainly due to increases in livestock population and milk production.

3.0. Options for Reducing Methane Emissions

One of the key elements of the U.S. Climate Change Action Plan (CCAP) is the implementation of cost-effective reductions of methane emissions through voluntary industry actions.[3] Because methane is a valuable energy resource, recovering methane that normally would be emitted into the atmosphere and using it for fuel reduces greenhouse gas emissions. The methane saved from these voluntary actions often pays for the costs of recovery and also can be cost-effective even without accounting for the broader social benefits of reducing greenhouse gases (GHG).

Beginning in the early 1990s, EPA launched five voluntary programs to promote cost-effective methane emission reductions:

- AgSTAR Program – works with livestock producers to encourage methane recovery from animal waste;
- Coalbed Methane Outreach Program (CMOP) – works with the coal and natural gas industries to collect and use methane that is released during mining;
- Landfill Methane Outreach Program (LMOP) – works with states, municipalities, utilities, and the landfill gas-to-energy industry to collect and use methane from landfills;
- Natural Gas STAR Program – works with the companies that produce, transmit, and distribute natural gas to reduce leaks and losses of methane; and
- Ruminant Livestock Efficiency Program (RLEP) – works with livestock producers to improve animal nutrition and management, thereby boosting animal productivity and cutting methane emissions.

Under these voluntary programs, industry partners voluntarily undertake cost-effective efforts to reduce methane emissions. EPA works with partners to quantify the results of their actions and account for reductions in historical methane emission estimates. One of the principal benefits of these voluntary programs is the sharing of information between government and industry and within industry on emissions, and emission reduction opportunities and associated costs. These programs have contributed significantly to EPA's understanding of the opportunities for emission reductions.

Many of these opportunities involve the recovery of methane emissions and use of the methane as fuel for electricity generation, on-site heat uses, or off-site sales of methane. These actions represent key opportunities for reducing methane emissions from landfills, coal mines, and livestock manure management. Other options may include oxidizing or burning the methane emissions. Catalytic oxidation is a new technology potentially applicable at coal mines; flaring is an option available at landfills and other sites.

The natural gas industry offers the most robust array of emission reduction options. The Natural Gas STAR Program has identified a number of best management practices for reducing leaks and avoiding venting of methane. In addition, partners in the program have employed a number of other strategies for reducing emissions. These strategies are described in the chapter on natural gas systems.

Conversely, few technology-specific reduction options have yet been identified for the ruminant livestock industry, where methane production is a natural byproduct of enteric fermentation. The principal options are improving the efficiency of feedlot operations and animal feeds for ruminant livestock. Better feeds and animal management can increase yields of meat and dairy products relative to methane production.

A principal benefit of the various voluntary programs is abundant information developed on the efficacy of the emission reduction options and the costs of implementing these options. EPA uses this information to estimate the costs of reducing emissions. Partners in the various voluntary programs are already undertaking emission reduction efforts because they have been found to be cost-effective. While some of the emission reduction options are cost-effective in some settings, they are not in others, e.g., methane recovery and use may be more cost-effective at large coal mines and landfills than at small ones. In the next section the economics of decision making in the implementation of reduction options is discussed.

4.0. Economic Analysis of Reducing U.S. Methane Emissions

This report presents the results of extensive benefit-cost analyses conducted on the opportunities (technologies and management practices) to reduce methane emissions from four of the five major U.S. sources: landfills, natural gas systems, coal mining, and livestock manure. The analyses are conducted for the years 2000, 2010, and 2020. EPA selected these sources because well-characterized opportunities exist for cost-effective emission reductions. The results are in terms of abated methane (emission reductions) that can be achieved at various values of methane. The total value of methane is the sum of its value as a source of energy and as an emission reduction of a GHG.

Methane has a value as a source of energy since it is the principal component of natural gas. Therefore, avoided methane emissions in natural gas systems are valued in terms of dollars per million British thermal units ($/MMBtu). Similarly, methane also can be combusted to generate electricity and is valued in dollars per kilowatt-hour ($/kWh). The value of potential methane emission reductions is calculated relative to carbon equivalent units using methane's 100-year global warming potential (GWP) of 21 (IPCC, 1996a). The value of abated methane, as well as other GHGs, can thus be stated in terms of dollars per metric ton of carbon equivalent ($/TCE). Throughout the analysis, energy market prices are aligned to $0/TCE. This value represents a scenario where no additional price signals from GHG abatement values exist to motivate emission reductions; all reductions are due to responses to market prices for natural gas. As a value is placed on GHG reductions in terms of $/TCE, these values are added to energy market prices and allow for additional emission reductions to clear the market.

A benefit-cost analysis is applied to the opportunities for emission reductions and is defined as:

- Benefits. Benefits are calculated from the amount of methane saved by implementing the options multiplied by the value of the methane saved as its use as an energy resource; plus the value of methane as an emission reduction of a GHG, if available;
- Costs (including capital expenditures and operation and maintenance expenses). The costs of implementing specific reduction options are estimated for four of the five major anthropogenic sources. The applied discount rates are particular to each source-specific analysis and set at eight percent for the aggregate analysis.4 In the source-specific analyses, different discount rates are used to determine cost-effective reductions.

Because nearly all of the technologies and practices for reducing methane emissions produce or save energy, energy prices are a key driver of the cost analyses. The value of the energy produced or saved offsets to various degrees the capital and operating costs of reducing the emissions. Higher energy prices offset a larger portion of these costs, and in some cases make the technologies and practices profitable.[5]

In the source-specific analyses, energy market prices, in 1996 U.S. dollars, are used to establish whether an option is cost-effective. These prices are established based on the following approaches:

- For landfills, both electricity and natural gas prices are used in the analysis since landfills sell gas directly to consumers or use the recovered gas to generate electricity. For electricity prices, the analysis uses an estimated price of $0.04/kWh to represent the value of electricity close to distribution systems and receiving a renewable energy premium. For natural gas, the price used is $2.74/MMBtu. In this case, the analysis uses the average industrial gas price discounted by 20 percent to adjust for the lower Btu content of landfill gas (EIA, 1997).
- Coal mine methane is sold as natural gas to interstate pipelines, used to generate electricity, or used on-site. For natural gas, coal mine

methane is valued at \$2.53/MMBtu, which is the average delivered price for natural gas in Alabama, Indiana, Kentucky, and Ohio. The electricity generated from coal mines is valued at \$0.03/kWh to reflect the greater distance from distribution systems.

- The set of energy prices for natural gas systems depends on where the emissions are reduced. Production emission reductions are valued at the average wellhead price of \$2.17/MMBtu; transmission savings are valued at \$2.27/MMBtu; and distribution system savings are valued at \$3.27/MMBtu (EIA, 1997).
- Livestock manure methane is used to generate electricity for farm use and offset electricity consumption from a utility grid. The analysis uses \$0.09/kWh for dairy farms and \$0.07/kWh for swine farms. These prices are weighted averages of retail commercial electricity rates based on dairy and swine populations, respectively. The national average price was discounted by \$0.02/kWh to reflect the effects of interconnect and demand charges and other associated costs.

In order to incorporate methane emission reduction values into the analysis, various \$/TCE values are translated into equivalent electricity and gas prices using the heat rate of the engine-generator (for electricity), the energy value of methane (1,000 Btu/cubic foot), and a GWP of 21. See individual chapters for greater detail.

5.0. Achievable Emission Reductions and Composite Marginal Abatement Curve

The aggregate results of the analyses are presented in this section. Exhibit 1-5 shows estimated total U.S. reductions at various values for abated methane in \$/TCE. These reductions are the summation of source-specific results where different discount rates are applied to each source: 8 percent for landfills, 10 percent for livestock manure management, 15 percent for coal mining, and 20 percent for natural gas systems. For 2010, EPA estimates that up to 34.8 MMTCE (6.1 Tg) of reductions are possible at energy market prices or \$0/TCE. Consequently, methane emissions could be reduced below 1990 emissions of 169.9 MMTCE (29.7 Tg) if many of the identified opportunities are thoroughly implemented. At higher emission reduction values, more methane reductions could be achieved. For example, EPA's analysis indicates that with a value of $20/TCE for abated methane added to the energy market price, U.S. reductions could reach 50.3 MMTCE (8.8 Tg) in 2010.

Exhibit 1-6 presents EPA's aggregate U.S. methane marginal abatement curve (MAC) for 2010 which is calculated using a discount rate of eight percent equally applied to all sources in order to properly construct the curve.[4] The MAC illustrates the amount of reductions possible at various values for methane and is derived by rank ordering individual opportunities by cost per emission reduction amount (IPCC, 1996b). Any point along a MAC represents the marginal cost of abating an additional amount of methane. A complete picture is revealed when the prevailing market prices for energy and GHG reductions are applied to the MAC to show the amount of available emissions that clear the market. Any "below-the-line" reduction amounts, with respect to $0/TCE, illustrate this dual price-signal market, i.e., energy market prices and emission reduction values.

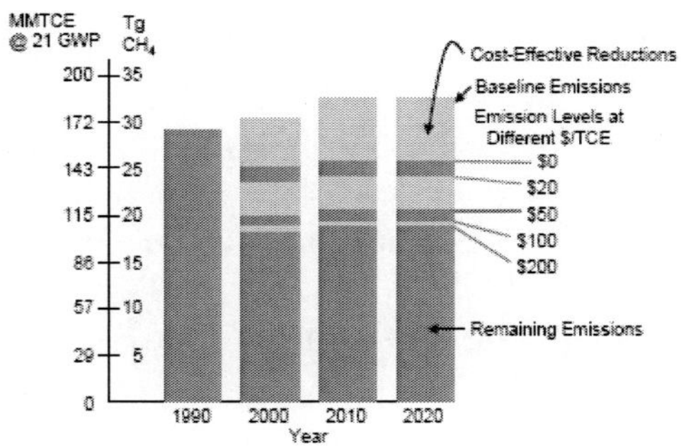

	2000	2010	2020
Baseline Emissions	173.9	186.0	183.7
Cumulative Reductions			
at $0/TCE	30.6	34.8	35.0
at $10/TCE	36.4	42.3	40.9
at $20/TCE	41.7	50.3	47.4
at $30/TCE	54.6	61.7	58.7
at $40/TCE	56.2	63.5	61.0
at $50/TCE	59.5	66.9	64.8
at $75/TCE	64.3	71.9	70.7
at $100/TCE	67.2	74.9	74.0
at $125/TCE	68.4	76.2	75.5
at $150/TCE	68.7	76.5	75.9
at $175/TCE	69.0	76.8	76.2
at $200/TCE	69.2	77.0	76.5
Remaining Emissions	104.7	108.9	107.2

Exhibit 1-5. U.S. Baseline Emissions and Potential Reductions (source-specific discount rates) (MMTCE).

The MAC illustrates the following key findings. First, substantial emission reductions, 36.8 MMTCE (6.4 Tg), can be cost-effectively achieved, that is, at energy market prices with no additional emissions reduction values or $0/TCE. Second, at $20/TCE and $50/TCE estimated reductions are 52.6 MMTCE (9.2 Tg) and 70.0 MMTCE (12.2 Tg), respectively. Third, at $100/TCE, achievable reductions are estimated at 75.5 MMTCE (13.2 Tg). Finally, above $100/TCE, the MAC becomes inelastic, that is, non-responsive to increasing methane values which indicates the limits of the options considered. At higher energy and emission reduction values, additional options, which have yet to be developed, will likely become available. By not estimating potential, future higher-cost options, this analysis under-estimates the ability to reduce emissions at higher values for abated methane.

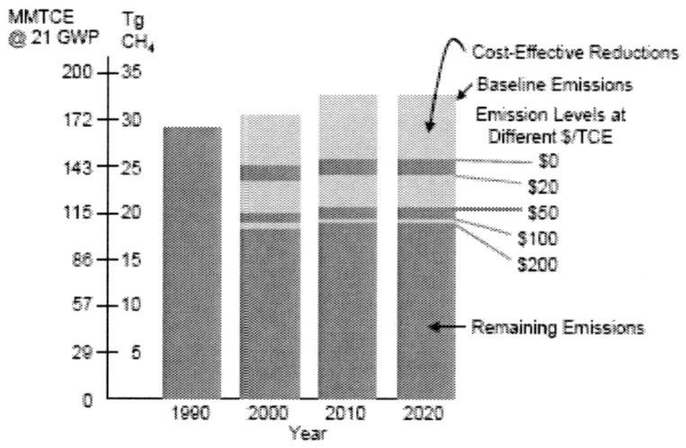

	2000	2010	2020
Baseline Emissions	173.9	186.0	183.7
Cumulative Reductions			
at $0/TCE	30.6	34.8	35.0
at $10/TCE	36.4	42.3	40.9
at $20/TCE	41.7	50.3	47.4
at $30/TCE	54.6	61.7	58.7
at $40/TCE	56.2	63.5	61.0
at $50/TCE	59.5	66.9	64.8
at $75/TCE	64.3	71.9	70.7
at $100/TCE	67.2	74.9	74.0
at $125/TCE	68.4	76.2	75.5
at $150/TCE	68.7	76.5	75.9
at $175/TCE	69.0	76.8	76.2
at $200/TCE	69.2	77.0	76.5
Remaining Emissions	104.7	108.9	107.2

Exhibit 1-6. Marginal Abatement Curve for U.S. Methane Emissions in 2010 (at an 8 percent discount rate).

The MAC is based on approximately 160 observations. These results are from the benefit-cost analyses conducted on the identified opportunities to abate methane emissions.

An analytic approximation of the MAC is calculated in order to make these results useful to larger economic models concerned with GHG reduction costs. The estimated relationship is obtained by using an exponential trendline, expressing the relationship between methane values/abatement costs and the quantity of abated methane.[6] This function is described as: $/TCE = 30 exp [45/(102 – MMTCE)]-60.

Exhibit 1-7 illustrates the relative contribution of each of the sources to reducing methane emissions. Of the four sources, landfills contribute the most to the emission reductions, i.e., over one-quarter of the reductions. Coal mining and natural gas systems each account for about one-quarter of total emission reductions. Livestock manure contributes up to about one-fifth of the reductions, primarily at higher energy prices and emission reduction values. Several key aspects of the analysis are highlighted below:

- The methane recovery efficiency at landfills is estimated at 75 percent for all landfills and is assumed to remain constant. Below $0/TCE, using the recovered methane directly in boilers or similar equipment is more cost-effective than producing electricity in most cases.
- Because of the diverse sources of methane emissions from natural gas systems, a large number of technologies and practices are evaluated. Among the options evaluated, replacing high-bleed pneumatic devices and techniques for reducing emissions from compressor stations are the most significant in terms of cost-effective emission reductions.

- The coal mine methane analysis includes a catalytic oxidation technology for recovering heat energy from the low concentration of methane in coal mine ventilation air. This technology becomes profitable at approximately $30/TCE, leading to substantial emission reductions from underground mining. Below this value, methane recovery is the primary method of reducing emissions.
- The principal methods for reducing methane emissions from livestock manure are to collect and combust the methane that would otherwise be emitted from liquid manure management systems. Anaerobic digester technologies, the principal technology evaluated, produce multiple benefits, including odor reduction at swine farms as well as producing energy for on-farm use.

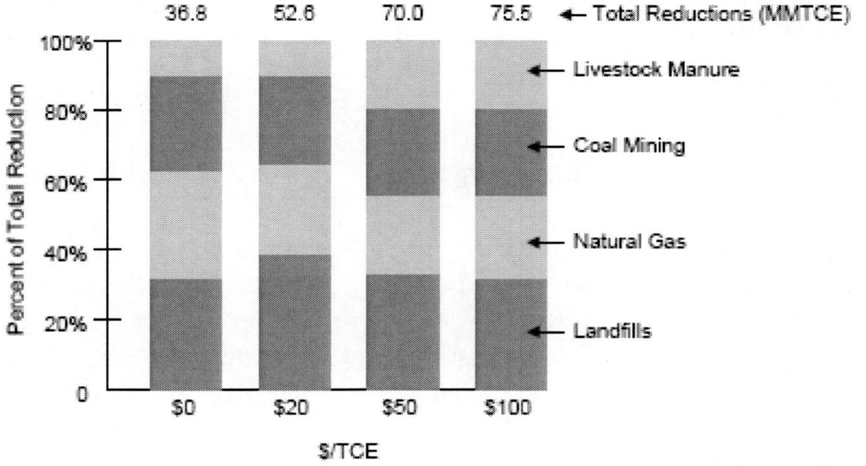

Exhibit 1-7. Portion of Emission Reductions from Each Source in 2010 (at an 8 percent discount rate) (MMTCE).

6.0. Significance of This Analysis

To date, most economic analyses of GHG reduction opportunities have focused on energy-related carbon emissions since CO_2 currently accounts for about 82 percent of the total U.S. emissions (weighted by 100- year global warming potentials) (EPA, 1999). The analyses provided in this report can be integrated with CO_2 economic analyses to provide a broader understanding of reducing the total cost of achieving GHG emission reductions. Recent comprehensive studies by the Joint Program on the Science and Policy of Global Change, Massachusetts Institute of Technology (Reilly, 1999) and the Australian Bureau for Agricultural and Resource Economics (Brown, 1999) show that a multi-gas mitigation strategy can reduce the costs of achieving GHG emission reductions. Both of these studies utilized EPA's preliminary cost analysis on methane reductions (EPA, 1998).

The economic benefits of pursuing a mitigation strategy that includes methane is shown in Exhibit 1-8. Illustrative MACs are presented for methane (CH_4), CO_2, and for the summation of the two showing additional emission reductions with increases in $/TCE. Given a reduction target, A*, for both gases, the total cost of achieving that target is lower if available methane reductions are included than if only CO_2 reductions are made. At increasing values for emission reductions, more costly CO_2 reductions can be substituted by lower cost methane reductions, when available, thereby lowering the marginal cost, shown as the movement from P to P·, and decreasing the total cost (the integral or area under the curve).

7.0. Background to This Report

EPA's first major report on methane appeared in 1993 as *Anthropogenic Methane Emissions in the United States, Estimates for 1990, Report to Congress* (1993a). This report was the first effort to increase general knowledge about methane emissions by presenting a detailed and comprehensive treatment of the sources of methane emissions as part of the effort to quantify these emissions. Following this report, EPA published Opportunities to Reduce Anthropogenic Methane Emissions in the United States (EPA, 1993b). For all major sources of methane emissions – landfills, natural gas systems, coal mines, livestock manure, and livestock enteric fermentation – this report described the technologies available that could reduce emissions. Using these technologies, the report estimated the amount of emission reductions that would be technically feasible and the amount of emission reductions that would be economically justified. The latter included taking into account the value of methane (as a fuel) as well as a value for emission reductions.

Since the publication of these reports, EPA has sponsored additional work in the estimation of baseline emissions and the costs of emission reductions. These efforts include, for example, a 15-volume report on Methane Emissions from Natural Gas Systems cosponsored with the Gas Research Institute (EPA/GRI, 1996).

The information from the various voluntary programs in addition to other research was used extensively in the EPA's Costs of Reducing Methane Emissions in the United States, Preliminary Report (EPA, 1998). This report first developed the overall approach for estimating the cost of emission reductions and was reviewed by a number of industry and source experts. Their subsequent recommendations as well as other improvements have been incorporated into the current document.

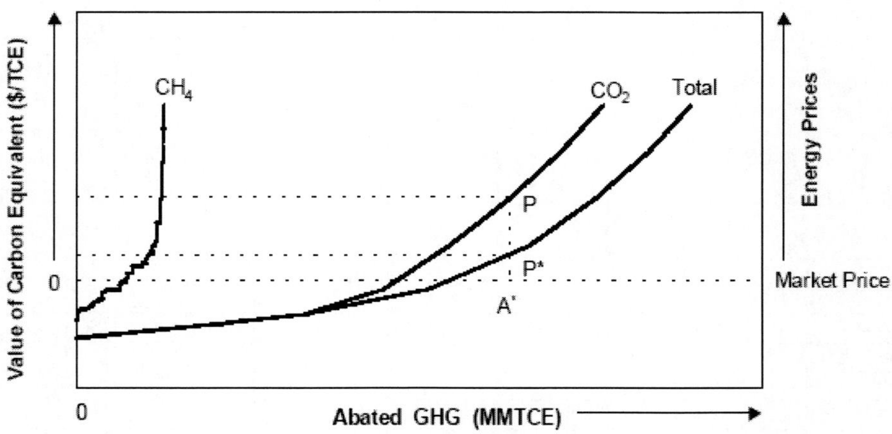

Exhibit 1-8. Illustrative MACs for Methane and Carbon Dioxide.

APPENDIX I. SUPPORTING MATERIAL FOR COMPOSITE MARGINAL ABATEMENT CURVE

This appendix presents the data EPA used to develop the composite marginal abatement curve (MAC). The first section summarizes the incremental emissions reductions associated with each source, i.e., landfills, natural gas systems, coal mining, and livestock manure. The second section presents the approach to fit an equation to the MAC data.

I.1. Estimates for Composite Marginal Abatement Curve

This section presents estimates of the incremental emission reductions for each combination of carbon equivalent value and methane source. Exhibit I-1 presents these estimates. The exhibit also includes the cumulative emission reductions. These cumulative emission reductions form the composite MAC for 2010.

Exhibit I-1. Composite Marginal Abatement Curve Schedule of Options for 2010

Value of Carbon Equiva-lent $/TCE	Incre-mental Reduc-tions (MMTCE)	Source	Cumulative Reductions (MMTCE)	Value of Carbon Equivalent $/TCE	Incremental Reductions (MMTCE)	Source	Cumulative Reductions (MMTCE)
($30.00)	0.29	Manure-Dairy	0.29	($16.32)	0.25	Coal	13.91
($30.00)	1.23	Manure-Swine	1.52	($16.00)	0.98	Natural Gas	14.89
($23.72)	0.45	Natural Gas	1.98	($15.74)	0.19	Natural Gas	15.08
($23.62)	0.23	Natural Gas	2.20	($15.67)	0.09	Natural Gas	15.17
($23.24)	0.64	Natural Gas	2.85	($15.11)	0.73	Natural Gas	15.89
($23.01)	0.12	Natural Gas	2.96	($14.45)	0.05	Natural Gas	15.95
($22.95)	0.24	Natural Gas	3.20	($14.41)	0.35	Natural Gas	16.30
($20.85)	0.32	Natural Gas	3.52	($14.14)	0.41	Coal	16.71
($20.00)	0.77	Manure-Dairy	4.29	($14.02)	0.14	Natural Gas	16.86
($19.86)	0.33	Natural Gas	4.62	($13.41)	0.29	Coal	17.15
($19.77)	0.42	Natural Gas	5.04	($12.17)	0.90	Natural Gas	18.04
($19.51)	0.87	Coal	5.91	($11.78)	0.31	Coal	18.35
($19.32)	1.63	Natural Gas	7.54	($11.50)	0.26	Coal	18.61
($19.18)	0.01	Natural Gas	7.55	($11.32)	0.41	Coal	19.02
($19.14)	0.79	Coal	8.34	($11.01)	0.20	Natural Gas	19.22
($19.13)	0.59	Natural Gas	8.93	($10.65)	0.04	Natural Gas	19.27
($18.96)	1.63	Coal	10.55	($10.59)	0.16	Coal	19.43
($18.87)	0.77	Coal	11.32	($10.50)	0.42	Coal	19.84
($18.69)	0.57	Coal	11.89	($10.39)	0.65	Natural Gas	20.49
($18.42)	0.48	Coal	12.37	($10.28)	0.02	Natural Gas	20.52
($16.86)	0.39	Natural Gas	12.76	($10.00)	0.62	Manure-Dairy	21.14
($16.70)	0.43	Natural Gas	13.20	($9.51)	0.04	Natural Gas	21.18
$16.41)	0.47	Coal	13.67	($9.23)	0.19	Coal	21.37
($9.16)	0.56	Natural Gas	21.93	$12.41	0.09	Coal	43.32
($7.87)	0.47	Coal	22.40	$12.78	0.11	Coal	43.43
($7.68)	0.38	Coal	22.78	$12.87	0.09	Coal	43.52

Value of Carbon Equivalent $/TCE	Incremental Reductions (MMTCE)	Source	Cumulative Reductions (MMTCE)	Value of Carbon Equivalent $/TCE	Incremental Reductions (MMTCE)	Source	Cumulative Reductions (MMTCE)
($7.50)	0.39	Natural Gas	23.17	$14.32	0.03	Coal	43.55
($6.92)	0.06	Natural Gas	23.24	$15.60	0.16	Coal	43.71
($6.77)	0.33	Coal	23.57	$16.23	0.07	Coal	43.78
($6.50)	0.09	Coal	23.66	$16.51	0.14	Coal	43.92
($6.23)	0.22	Coal	23.88	$16.78	0.11	Coal	44.03
($4.77)	0.34	Coal	24.21	$16.87	0.03	Coal	44.06
($3.80)	0.01	Natural Gas	24.22	$17.51	0.09	Coal	44.15
($3.23)	0.20	Coal	24.42	$18.42	0.06	Coal	44.21
($2.50)	0.14	Coal	24.56	$18.71	0.06	Natural Gas	44.27
($1.61)	0.01	Natural Gas	24.57	$18.84	0.35	Natural Gas	44.63
($1.41)	0.17	Coal	24.74	$18.84	0.22	Natural Gas	44.84
($1.32)	0.07	Coal	24.81	$19.06	0.14	Natural Gas	44.98
($0.86)	0.27	Coal	25.07	$19.69	0.06	Coal	45.04
($0.82)	0.60	Natural Gas	25.67	$20.00	0.20	Manure-Dairy	45.24
($0.59)	0.03	Coal	25.70	$20.00	1.54	Manure-Swine	46.78
($0.05)	0.10	Coal	25.80	$20.00	5.79	Landfills	52.57
$0.00	0.50	Manure-Dairy	26.30	$21.14	0.04	Coal	52.62
$0.00	10.55	Landfills	36.85	$21.51	0.02	Coal	52.63
$0.41	0.06	Coal	36.91	$22.87	0.07	Coal	52.70
$0.95	0.16	Coal	37.07	$23.96	0.05	Coal	52.75
$1.05	0.07	Coal	37.13	$24.51	0.03	Coal	52.77
$1.32	0.25	Coal	37.38	$24.65	0.00	Natural Gas	52.77
$2.05	0.15	Coal	37.53	$27.87	0.06	Coal	52.83
$3.51	0.15	Natural Gas	37.68	$29.70	6.28	Coal	59.10
$4.96	0.02	Coal	37.70	$30.00	0.18	Manure-Dairy	59.28
$5.23	0.24	Coal	37.94	$30.00	2.28	Manure-Swine	61.57
$5.25	0.02	Natural Gas	37.96	$30.00	1.22	Landfills	62.79
$6.45	0.14	Natural Gas	38.10	$31.59	0.51	Natural Gas	63.30
$6.58	0.04	Natural Gas	38.14	$35.52	0.77	Natural Gas	64.07
$6.60	0.10	Natural Gas	38.24	$35.52	0.00	Natural Gas	64.07
$7.19	0.03	Natural Gas	38.27	$38.14	0.87	Natural Gas	64.94
$7.62	0.21	Natural Gas	38.47	$38.60	0.42	Natural Gas	65.36
$9.32	0.18	Coal	38.65	$39.77	0.00	Natural Gas	65.36
$9.59	0.03	Coal	38.68	$40.00	0.16	Manure-Dairy	65.52
$10.00	0.31	Manure-Dairy	39.00	$40.00	1.45	Manure-Swine	66.97
$10.00	0.12	Manure-Swine	39.11	$40.00	0.29	Landfills	67.26
$10.00	3.89	Landfills	43.01	$40.88	0.00	Natural Gas	67.26
$11.23	0.03	Coal	43.04	$45.21	0.94	Natural Gas	68.20
$11.41	0.04	Coal	43.08	$47.09	0.32	Natural Gas	68.52
$11.69	0.07	Coal	43.14	$47.54	0.02	Natural Gas	68.54

Exhibit I-1. (Continued)

Value of Carbon Equiva-lent $/TCE	Incre-mental Reduc-tions (MMTCE)	Source	Cumulative Reductions (MMTCE)	Value of Carbon Equivalent $/TCE	Incremental Reductions (MMTCE)	Source	Cumulative Reductions (MMTCE)
$12.04	0.00	Natural Gas	43.14	$50.00	0.16	Manure-Dairy	68.70
$12.14	0.09	Coal	43.23	$50.00	1.18	Manure-Swine	69.88
$50.00	0.11	Landfills	69.98	$100.00	0.02	Landfills	75.54
$52.10	0.67	Natural Gas	70.65	$113.08	0.12	Natural Gas	75.66
$56.12	0.56	Natural Gas	71.22	$116.47	0.45	Natural Gas	76.10
$65.77	0.00	Natural Gas	71.22	$125.00	0.30	Manure-Dairy	76.41
$75.00	0.42	Manure-Dairy	71.63	$125.00	0.08	Manure-Swine	76.49
$75.00	2.77	Manure-Swine	74.40	$140.29	0.01	Natural Gas	76.50
$75.00	0.05	Landfills	74.45	$150.00	0.27	Manure-Dairy	76.77
$76.24	0.08	Natural Gas	74.53	$166.22	0.03	Natural Gas	76.80
$95.34	0.21	Natural Gas	74.74	$175.00	0.23	Manure-Dairy	77.03
$95.47	0.00	Natural Gas	74.74	$188.35	0.07	Natural Gas	77.10
$100.00	0.38	Manure-Dairy	75.12	$200.00	0.19	Manure-Dairy	77.29
$100.00	0.40	Manure-Swine	75.52				

I.2. Equation for Composite Marginal Abatement Curve

The relationship between the additional value of carbon equivalent ($/TCE) and the cumulative emission reductions, i.e., abated methane in MMTCE is shown in Exhibit II-2. The cumulative emission reductions increases relatively slowly as a function of the value of carbon equivalent. As the cumulative emission reductions reach about 75 MMTCE, the reduction plateau and cannot be further abated at higher $/TCE values. In order to represent the steepness of the curve at values close to 75 MMTCE, EPA determined a best-fit curve based on the data points. This equation is defined by:

$$y = \text{parameter}_1 * \exp\,[\text{parameter}_2 \,/\, (\max - x)]\ \text{offset}$$

where:

y = additional value of carbon equivalent ($/TCE)

x = cumulative emission reductions (MMTCE)

parameter $_1$, parameter $_2$, offset, and max = determined parameters

All values of x, i.e., cumulative emission reductions, must be less than the value of max. This curve has the property that as the x value increases to the value of max, the y value will tend to infinity, so the curve will approximate the steep rise at the maximum x value.

EPA used the method of least squares to find the best fitting curve. This method estimates the parameters by minimizing the mean square error (MSE), i.e., the average squared difference between the actual and fitted values of y: MSE = (actual y fitted $y)^2$ / n, where n is the number of pairs, i.e., 159 pairs of abated methane and additional value of carbon equivalent. The minimum MSE is 68.6. The fitted parameters are:

- offset = 60
- parameter$_1$ = 30
- parameter$_2$ = 45
- max = 102

The resulting equation is given by:

$y = 30 * \exp [45/(102\text{-}x)] - 60$

The squared correlation coefficient (R squared) between the actual and predicted values of y is 0.95, showing a reasonably good fit on a scale of zero to one, one being a perfect fit. Although the model was fitted using the method of least squares, the optimum least squares solution for this problem is also the solution with the maximum possible R squared. Exhibit II-2 presents the 159 data points and the fitted curve.

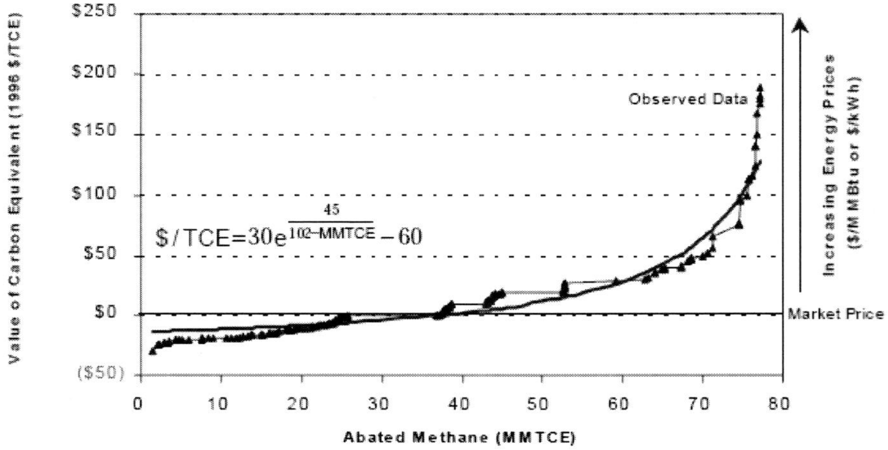

Exhibit II-2. Marginal Abatement Curve for U.S. Methane Emissions in 2010.

REFERENCES

Boden, T.A., D.P. Kaiser, R.J. Sepanski, and F.W. Stoss. 1994. *Trends '93: A Compendium of Data on Global Change*. World Data Center for Atmospheric Trace Gases, Carbon Dioxide Information Analysis Center, Environmental Sciences Division, Oak Ridge National Laboratory, Oak Ridge, TN, ORNL/CDIAC-65, ESD Publication No. 4195.

Brown, Stephen, Darren Kennedy, Cain Polidano, Kate Woffenden, Guy Jakeman, Brett Graham, Frank Jotzo, and Brian S. Fisher. 1999. "Assessing the economic impacts of the Kyoto Protocol Implications of accounting for the three major greenhouse gases." Australian Bureau for Agricultural and Resource Economics (ABARE). ABARE Research Report 99.6, Canberra, Australia, May 1999. (Available on the Internet at http://www. abare.gov.au/pdf/RR99.6pdf.)

Dlugokencky, E.J., K.A. Masarie, P.M. Lang, and P.P. Tans. 1998. "Continuing Decline in the Growth Rate of the Atmospheric Methane Burden," *Nature*, v. 393, 4 June 1998.

DOE. 1998. Statement of Robert S. Kripowicz, Principal Deputy Assistant Secretary for Fossil Energy, U.S. Department of Energy Before the Subcommittee on Energy, Research, Development, Production, and Regulation, U.S. Senate, Washington, DC, 21 May 1998.

DOS. 1997. *Climate Action Report: 1997. Submission of the United States of America Under the United Nations Framework Convention of Climate Change.* Bureau of Oceans and International Environmental Scientific Affairs, Office of Global Climate Change, U.S. Department of State, Washington, DC, DOS 10496. (Available on the Internet at http://www.state.gov/www/global/oes/97climate_report/index.html.)

EIA. 1997. *Natural Gas Annual 1996.* Office of Oil and Gas, Energy Information Administration, U.S. Department of Energy, Washington, DC, DOE/EIA-0540(96). (Available on the Internet at http://www.eia.doe.gov/oil_gas/ natural_gas/nat_frame.html.)

EIA. 1998. *Annual Energy Outlook (AEO) 1998.* Reference Case Forecast, Energy Information Administration, U.S. Department of Energy, Washington, DC.

EPA. 1993a. *Anthropogenic Methane Emissions in the United States: Estimates for 1990, Report to Congress*, Atmospheric Pollution Prevention Division, Office of Air and Radiation, U.S. Environmental Protection Agency, Washington, DC, EPA 430-R-93-003. (Available on the Internet at http.//www.epa.gov/ghg info/reports.htm.)

EPA. 1993b. *Current and Future Methane Emissions from Natural Sources, Report to Congress.* Atmospheric Pollution Prevention Division, Office of Air and Radiation, U.S. Environmental Protection Agency, Washington, DC, EPA 430-R-93-011. (Available on the Internet at http://www.epa.gov/ghginfo/reports.htm.)

EPA. 1994. *International Anthropogenic Methane Emissions: Estimates for 1990, Report to Congress.* Atmospheric Pollution Prevention Division, Office of Air and Radiation, U.S. Environmental Protection Agency, Washington, DC, EPA 230-R-93-010.

EPA. 1998. *Costs of Reducing Methane Emissions in the United States, Preliminary Report, Draft.* Methane and Utilities Branch, Atmospheric Pollution Prevention Division, Office of Air and Radiation, U.S. Environmental Protection Agency, Washington, DC.

EPA. 1999. *Inventory of Greenhouse Gas Emissions and Sinks 1990-1997.* Office of Policy, Planning, and Evaluation, U.S. Environmental Protection Agency, Washington, DC; EPA 236-R-99-003. (Available on the Internet at http://www.epa.gov/globalwarming/inventory/1999-inv.html.)

EPA/GRI. 1996. *Methane Emissions from the Natural Gas Industry, Volume 1: Executive Summary.* Prepared by M. Harrison, T. Shires, J. Wessels, and R. Cowgill, eds., Radian International LLC for National Risk Management Research Laboratory, Air Pollution Prevention and Control

Division, U.S. Environmental Protection Agency and Gas Research Institute, Research Triangle Park, NC, EPA-600-R-96-080a.

IPCC. 1990. *Climate Change: The IPCC Scientific Assessment.* Intergovernmental Panel on Climate Change (IPCC), Cambridge University Press, Cambridge, United Kingdom.

IPCC. 1995. *Climate Change 1994: Radiative Forcing of Climate Change.* Intergovernmental Panel on Climate Change (IPCC), Cambridge University Press, Cambridge, United Kingdom.

IPCC. 1996a. *Climate Change 1995: The Science of Climate Change.* Intergovernmental Panel on Climate Change (IPCC), Cambridge University Press, Cambridge, United Kingdom.

IPCC. 1996b. *Climate Change 1995: Economic and Social Dimension of Climate Change.* Intergovernmental Panel on Climate Change (IPCC), Cambridge University Press, Cambridge, United Kingdom.

Reilly, I., R.G. Prinn, J. Uarnisch, J. Fitzmaurice, H.D. Jacoby, D. Kickligher, P.H. Stone, A.P. Sokolov, and C. Wang. 1999. *Multi-Gas Assessment of the Kyoto Protocol, Report No. 45*, MIT Joint Program on the Science and Policy of Global Change, Boston, MA, January 1999.(Available on the Internet at http://web.mit.edu/globalchange/www/rp145.html.)

References for Executive Summary

Dlugokencky, E.J., K.A. Masarie, P.M. Lang, and P.P. Tans. 1998. "Continuing Decline in the Growth Rate of the Atmospheric Methane Burden," *Nature*, v. 393, 4 June 1998.

EPA. 1999. *Inventory of Greenhouse Gas Emissions and Sinks* 1990-1997. Office of Policy, Planning, and Evaluation, U.S. Environmental Protection Agency, Washington, DC. (Available on the Internet at http://www.epa.gov/globalwarming/inventory/index.html.)

Explanatory Notes

[1] The enhanced greenhouse effect is the concept that the natural greenhouse effect has been enhanced by anthropogenic emissions of greenhouse gases. Increased concentrations of carbon dioxide, methane, and nitrous oxide, CFCs, HFCs, PFCs, SF6, and other photochemically important gases caused by human activities such as fossil fuel consumption, trap more infra-red radiation, thereby exerting a warming influence on climate. Exhibit 1-1, which illustrates relative contributions to the enhanced greenhouse

effect by gas, is based on the increase in atmospheric concentrations at each gas between pre-industrial times and 1992. This exhibit does not include methane's indirect effect of tropospheric ozone and stratospheric water vapor production, which are estimated to be equivalent to about 25 percent of the direct effects.

[2] Microbial communities in upper soils constitute a much smaller methane sink.

[3] The U.S. CCAP was initiated in 1993 and designed to reduce U.S. emissions of greenhouse gases. CCAP Programs promote actions that are both cost-effective for individual private sector participants as well as beneficial to the environment.

[4] In the construction of a national or aggregate marginal abatement curve, a single discount rate is applied to all sources in order to equally evaluate various options. Given a particular value for abated methane, all options up to and including that value can be cost-effectively implemented. An eight percent discount rate, the lowest in the range of the source-specific rates (8 to 20 percent), is used since it is closer to social discount rates employed in national level analyses. The results from the single, eight percent discount rate analysis are slightly higher than the results where source-specific discount rates are used because a lower discount rate reduces project costs enabling additional reductions.

[5] The effects of energy price changes are analyzed only from the revenue side and do not consider effects to capital and O&M expenses. Therefore, the projected methane reductions may be overestimated for increases and underestimated for decreases to energy prices.

[6] For the estimated relationship, $\$/TCE = 30 \exp [45/(102 - MMTCE)] - 60$, the regression analysis yielded an R2 of 0.95. Conversely, the relationship also can be expressed in standard economic terms as the quantity of abated methane as a function of price ($/TCE): $MMTCE = 102 - 45/\ln [(\$/TCE+60)/30]$.

End Note for Executive Summary

[1] In the construction of a national or aggregate marginal abatement curve, a single discount rate is applied to all sources in order to equally evaluate various options. Given a particular value for abated methane, all options up to and including that value can be cost-effectively implemented. An eight percent discount rate, the lowest in the range of the source-specific rates (8 to 20 percent), is used since it is closer to social discount rates employed in national level analyses. The results from the single, eight percent discount rate analysis are slightly higher than the results where source-specific discount rates are used because a lower discount rate reduces project costs enabling additional reductions.

INDEX

C

D

E

I

K

L

M

Q

R

V

W

Y

Z